MELODY [illegible]

date production and protection

OVERLEAF: Hand pollinating

FAO PLANT PRODUCTION AND PROTECTION PAPER 35

date production and protection

with special reference to north africa and the near east

prepared by
horticultural crops group
plant production and protection division

based on the work of
dr. v.h.w. dowson

FOOD AND AGRICULTURE ORGANIZATION OF THE UNITED NATIONS
Rome 1982

The designations employed and the presentation of material in this publication do not imply the expression of any opinion whatsoever on the part of the Food and Agriculture Organization of the United Nations concerning the legal status of any country, territory, city or area or of its authorities, or concerning the delimitation of its frontiers or boundaries.

M

ISBN 92-5-101121-4

CONTENTS

LIST OF ILLUSTRATIONS

Page

Chapter X. Harvesting

Chapter XI. Pests and Diseases

Chapter XII. Improvement of Productivity

Note: The original drawings included in the above illustrations are all the work of Mrs. Joy Dowson.

LIST OF TABLES

FOREWORD AND ACKNOWLEDGEMENTS

FAO has taken a keen interest in date production for many years. As one of the important food crops of North Africa and the Near East, it has been the subject of enquiries from all governments in the region and FAO experts have, since 1954, worked in most of those countries and submitted reports on date production, protection, processing and marketing. Mr. V.H.W. Dowson, the author of the present work had a long experience in commercial date production in Iraq during the 1920's and 1930's. He later served as FAO date expert for nearly 20 years in Somalia, Libya and Iran and made numerous visits to all the main date-growing countries.

In 1962 he was joint author with A. Aten, of FAO's Land and Water Development Division, of a previous Agricultural Development Paper on the handling, processing and packing of dates. He took an active part in the first two international meetings on date production and processing organized by FAO in Tripoli in 1959 and in Baghdad, Iraq in 1965. These meetings have provided the main forum for date specialists in North Africa and the Near East in recent years and have been attended by delegates from governments throughout the region, by advisors from the Research Service and Department of Agriculture, U.S.A., and by specialists from the Institut Français de Recherches Fruitières Outre-Mer.

A third meeting was held in Baghdad in 1975 and a fourth meeting is planned within a few years. A Regional Palm and Date Research Centre was founded in 1978 in Baghdad and enjoys the participation of 18 FAO member countries.

The study on date production and protection now presented is addressed primarily to agronomists and horticulturists in the North African and Near East region. It is based mainly on the draft prepared by V.H.W. Dowson but owes much to the participation of his collaborators who supplies Chapters III, V, X, XI and XII and whose opinions are freely quoted throughout the text.

The late R.W. Nixon of the U.S. Date and Citrus Station, Indio, was a famous authority on date culture and date breeding. J.B. Carpenter, Superintendant of the U.S. Date and Citrus Station, Indio, H.S. Elmer and L.J. Klotz of the University of California, are eminent in their fields of pathology and entomology. P.F. Smead is a date production and handling specialist who has had a wide experience both as a private grower in California and as an FAO expert in Pakistan and Kenya.

Much of the technical comment here presented is based on the research work carried out at the U.S. Date and Citrus Station in Indio, California, on the annual reports of the Date Growers' Institute, Coachella, California, and on the papers presented at the three FAO Technical Conferences.

It is fully realised that prolonged work on local field trials in countries of the North African and Near Eastern region will be necessary before modern techniques can be applied with full efficiency. For this reason the authors have in many instances hesitated to give clear advice on production techniques but have relied on quotations from different authorities in date growing so that at least the various opinions may be noted and compared.

Sources have been documented in the list of references but for a full bibliography with over 3,000 titles, the reader is referred to the "Bibliography of the Date Palm" by V.H.W. Dowson (published by Field Research Projects, Coconut Grove, Miami, Florida 33133, U.S.A., 1976). This may be brought up to date to some degree by recourse to the recent reports of the Date Growers' Institute.

Attention is also drawn to another authoritative book on the subject, "The Date Palm" by Paul Popenoe, 1973, from the above publishers at Coconut Grove, Miami.

Acknowledgement is made of the technical editorial work of several horticulturists. The late F.W. Pansiot, former Chief of the Horticultural Branch of the Plant Production and Protection Division at FAO Headquarters provided the original inspiration for the work and his efforts were followed up by D.C. Giacometti, formerly Tropical Horticulturist in the Horticultural Branch. J.B. Carpenter gave much time to checking technical facts and references and L.B. Thrower, FAO Horticultural Consultant, re-arranged the text and produced the final draft.

It is hoped that the agricultural study now presented will be useful to all those interested in the modernisation and improvement of date growing in North Africa and the Near East and other areas with potential date-palm growing.

D.F.R. Bommer
Assistant Director-General
Agriculture Department

Table 1

THE NORTHERN LIMITS OF DATE PALM CULTIVATION FOR FRUIT IN THE NORTHERN HEMISPHERE OF THE OLD WORLD

Pakistan		
N.W.F. Province	Bannu District 33°N	Progress in the north blocked by the Himalayas. Protected on the west by the mountains of Waziristan and Baluchistan
Makran	Siahan Mountain Range 27°N	In the valleys
Iran		
Kermanshah	Qasr-i-Shirin 34°31'N	At the bottom of the northern face of the Zagros mountains
U.S.S.R.		
Turkmenistan	Kizyl Arvat about 39°N (on 1 000 000 sheet written Güzel Avrat)	Date palms introduced in 1935, gave their first crop in 1941. Now several thousand trees are bearing fruit (Evreinoff)
Iraq and Syria		
Along the northeast foothills	Taza Khurmatu 35°18'N	A few palms are found as far north as Erbil (36°15'N) where a large number were killed by the great frost of 1909-1910
Along the Tigris	Samarra 34°12'N	A few palms occur further upstream at Tekrit (34°35'N)
Along the Euphrates	Rawa 34°30'N	Some plantings have been reported further north, in Syria. There are date gardens at Abu Kemal 34°27'N, which, although they lie to the south of Rawa, are 116 km upstream of that town and in Syria
Israel and Lebanon		
Capernaum	32°53'N	An old planting near the lake

Table 1 (Continued)

On coast mostly south of Tripoli	34°26'N	Apparently seedlings. Some palms pollinated and fruit eaten fresh
In the great rift valley		Several old and many recent plantings from the Sea of Galilee to the Gulf of Aquaba
Cyprus and Turkey		
Nicosia	36°10'N	The best groves are round Nicosia, but the palm is cultivated in various places on the plains, and also north of Nicosia. Dates eaten fresh
Antalya	36°34'N	About 100 palms in the Citrus Research Station
Algeria		
El-Kantara	35°14'N	In a deep ravine at the southern limit of the Atlas mountains
Spain		
Elche	38°17'N	

Table 2

SOUTHERN LIMITS OF DATE PALM CULTIVATION IN THE NORTHERN HEMISPHERE OF THE OLD WORLD

India		
Gujarat	23°N	Occasional date palms are found in many places in India and some of them sometimes fruit. Some have even fruited at Trichinopoly (10°50'N) but nowhere except in the northwest are they important
Pakistan		
Sind	Kotri 25°22'N	

Table 2 (Continued)

Arabian Peninsula

The southern coast line, whose most southerly point, to the west of Aden is at 12°36'N

Tanzania

[Tabora in southern hemisphere 5°S]

Somalia

Genale	1°47'N	South of Mogadiscio, a little inland

Djibouti

Hambali	11°30'N	Close to Djibouti

Ethiopia

Dirre Dawa	10°15'N	

Republic of Sudan

Kamlin	15°02'N	On the Nile, a little south of Khartoum. Dates eaten fresh, fruit often spoiled by summer rains

Cameroon

Rei Buba	8°40'N	In the Garua district

Republic of Chad

Lettire	13°40'N	

Republic of Niger

Guidimouni	13°45'N	80 km east of Zinder

Republic of Upper Volta

Dori	14°10'N	

Republic of Mali

Kolokani	13°20'N	

Republic of Senegal

Bakel	14°51'N	

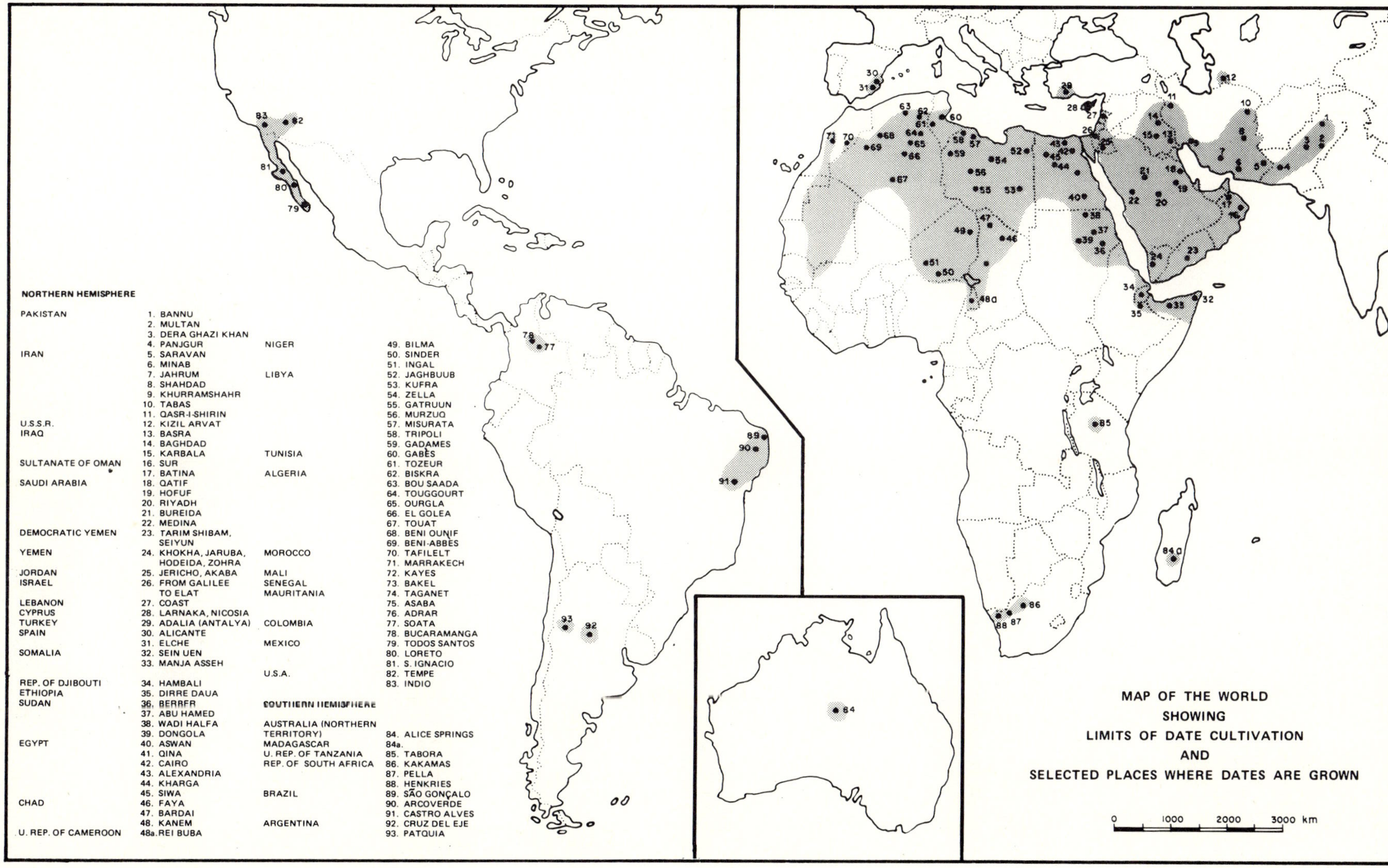
MAP OF THE WORLD
SHOWING
LIMITS OF DATE CULTIVATION
AND
SELECTED PLACES WHERE DATES ARE GROWN
0 1000 2000 3000 km
NORTHERN HEMISPHERE
PAKISTAN 1. BANNU
2. MULTAN
3. DERA GHAZI KHAN
4. PANJGUR
IRAN 5. SARAVAN
6. MINAB
7. JAHRUM
8. SHAHDAD
9. KHURRAMSHAHR
10. TABAS
11. QASR-I-SHIRIN
U.S.S.R. 12. KIZIL ARVAT
IRAQ 13. BASRA
14. BAGHDAD
15. KARBALA
SULTANATE OF OMAN 16. SUR
17. BATINA
SAUDI ARABIA 18. QATIF
19. HOFUF
20. RIYADH
21. BUREIDA
22. MEDINA
DEMOCRATIC YEMEN 23. TARIM SHIBAM, SEIYUN
YEMEN 24. KHOKHA, JARUBA, HODEIDA, ZOHRA
JORDAN 25. JERICHO, AKABA
ISRAEL 26. FROM GALILEE TO ELAT
LEBANON 27. COAST
CYPRUS 28. LARNAKA, NICOSIA
TURKEY 29. ADALIA (ANTALYA)
SPAIN 30. ALICANTE
31. ELCHE
SOMALIA 32. SEIN UEN
33. MANJA ASSEH
REP. OF DJIBOUTI 34. HAMBALI
ETHIOPIA 35. DIRRE DAUA
SUDAN 36. BERBER
37. ABU HAMED
38. WADI HALFA
39. DONGOLA
EGYPT 40. ASWAN
41. QINA
42. CAIRO
43. ALEXANDRIA
44. KHARGA
45. SIWA
CHAD 46. FAYA
47. BARDAI
48. KANEM
U. REP. OF CAMEROON 48a. REI BUBA
NIGER 49. BILMA
50. SINDER
51. INGAL
LIBYA 52. JAGHBUUB
53. KUFRA
54. ZELLA
55. GATRUUN
56. MURZUQ
57. MISURATA
58. TRIPOLI
59. GADAMES
TUNISIA 60. GABÈS
61. TOZEUR
ALGERIA 62. BISKRA
63. BOU SAADA
64. TOUGGOURT
65. OURGLA
66. EL GOLEA
67. TOUAT
68. BENI OUNIF
69. BENI-ABBÈS
MOROCCO 70. TAFILELT
71. MARRAKECH
MALI 72. KAYES
SENEGAL 73. BAKEL
MAURITANIA 74. TAGANET
75. ASABA
76. ADRAR
COLOMBIA 77. SOATA
78. BUCARAMANGA
MEXICO 79. TODOS SANTOS
80. LORETO
81. S. IGNACIO
U.S.A. 82. TEMPE
83. INDIO
SOUTHERN HEMISPHERE
AUSTRALIA (NORTHERN TERRITORY) 84. ALICE SPRINGS
MADAGASCAR 84a.
U. REP. OF TANZANIA 85. TABORA
REP. OF SOUTH AFRICA 86. KAKAMAS
87. PELLA
88. HENKRIES
BRAZIL 89. SÃO GONÇALO
90. ARCOVERDE
91. CASTRO ALVES
ARGENTINA 92. CRUZ DEL EJE
93. PATQUIA

CHAPTER 1

ORIGIN AND DISTRIBUTION

A. ORIGIN OF THE DATE PALM

The origin of the date palm (<u>Phoenix dactylifera</u> L.) is lost in antiquity. One of the world's earliest monuments, the temple of the moon god near Ur in Iraq, used the date palm in it's construction. It is thus certain that the date palm grew between 4 000 and 5 000 years ago. From what can be deduced from the temple itself and from the discovery of various objects of about the same date, it is known that it's builders were a highly developed people. It is therefore, at least probable that they practised agriculture and cultivated the date palm.

At a later period, that of the Assyrian empire, a stylized date palm is represented on the bas-reliefs at Nineveh. A winged figure is shown by it's side holding a cone. This relief was formerly considered to represent the pollintation of the date palm (Siret, 408); but E. Bonavia and C.J. Gadd (private communication to Mr. Nixon) have later suggested that a more likely explanation is that the fir-cone-like object held by the winged figure is an aspergillum, and that the palm is being sprinkled with holy water. The presence in the relief of a metal bucket, presumably to hold the holy water, lends support to this view.

A proof of the great antiquity of the date palm as a common object in Egypt, and the strong probability that it was cultivated at a very early stage in the civilization of that country are provided by it's being used as the symbol for a year in Egyptian hieroglyphics and it's frond as a symbol for a month.

Pliny the Elder, who was born in A.D. 23 or 24, has a chapter, No. 4 of Book 13, in his <u>Natural History</u> on the palms, which is mostly about the date palm. It reads much like an article on this crop in a modern encyclopaedia, except that he makes no mention of date culture in the United States.

B. DISTRIBUTION OF THE DATE PALM

1. <u>General</u>

The date palm needs a long, intensely hot summer, without rain or an excessively humid atmosphere for the five to seven months from pollination to harvest; but it also needs abundant irrigation. These requirements are found in places where there is water in the sub-tropical deserts, in both the northern and southern hemispheres, in North and South America, Africa, and Asia.

The home of the date palm is the lowland swept by the trade winds, from the Atlantic to the Indus, beyond the limit of the equatorial rain belt, roughly between 20°N and 35°N. East of Pakistan the monsoon rains make commercial date cultivation impossible.

For an apparent exception to the generalisation that the date palm needs a dry summer see Munier (300) and some of his other papers for an account of the date palm in tropical Africa.

Date Palms at Castello de Santa Barbara, Alicante, Spain. Dates ripen only imperfectly here.

Within comparatively recent times the date palm was introduced into the western parts of the other continents, to the Colorado Desert of North America, the Atacama Desert and other parts of South America, and the Kalahari Desert of South Africa, and into the great Central Desert of Australia. In North and South America the Spanish Conquistadores were responsible; the other introductions were more recent.

The old world date country stretches about 8 000 km from east to west and about 2 000 km from north to south, covering about 3% of the world's surface. Of these 16 000,000 square km, the area of cultivated land may be less than 1%. The dates possibly occupy 0.03% of the region.

The distribution of the date palm will be discussed from the two aspects of latitude and altitude.

2. Distribution according to Latitude

In the northern hemisphere the northern limit of the date palm is Venice, 44°24'N, but it does not fruit there. The most northerly commercial plantings of dates in the Old World date belt are the recent ones near Kizyl Arvat (Güzel Avrat) at 39°N on the edge of the Turkmenistan desert, where there are now several thousand palms, while at Bordighera on the Italian Riviera, about 44°N, there are the famous ornamental date palms, but these do not mature their fruit. The only commercial date fruit production in Europe is at Elche in Spain, where it was probably introduced by the Arabs, but possibly by the Phoenicians. Elche, close to Alicante, is at 38°17'N, and the dates attain only half maturity on the palm. It is not until Africa is reached, the Atlas mountains are crossed and the olives are left behind at El Kantara in the province of Constantine at 35°14'N that the date is found maturing properly.

The date palm then stretches southwards, where there is water, right across the Sahara, until it intermingles with the Doum palm at about 18°N. By the time the oil palm is reached at about 10°N, the date has almost disappeared, although there are a few date palms at Mogadiscio on the Somali coast at 2°N, and about 7 000 even as far south as Tabora at about 5°S, in Tanzania.

On the southern shore of the Arabian Peninsula, the Mahra coast and at Dhufar, there is a little mixed cultivation of the date and the coconut, but normally these two palms do not overlap.

The northern and southern limits of date palm cultivation for fruit in northern hemisphere of the Old World are summarized in Table (1) and (2).

3. Distribution according to Altitude

Moore (290) gives the average lapse rate in the atmosphere at 1°C in about 183 metres so that the temperatures 1 830 m up a mountain would be, other things being equal, 10°C colder that at it's foot. Of course there are complicating factors, such as the aspect (North or South) of sloping land and the tendency of cold air to collect at night in the bottom of the valley, but in general the distribution of date palms within the latitudes described is largely determined by the availability of water and the temperature limits imposed by

A part of the oasis of Shahdad, Iran, the home of Porkuu, the seedless date. Snow-clad mountains in the background. The Kuh-i-Sagueh peak in this range rises to 4 000 m. Kerman lies on the other side of the range.

altitude. It would seem from the following notes that in the date growing countries of the world, the palm flourishes within an altitude range of 1 892 m - from 392 m below to 1 500 m above sea level.

Pakistan

In Pakistan the date palm grows at an altitude of 527 m (at Rawalpindi). Further north it is grown commercially in Makran on the Siahan Mountains at over 600 m, and probably, on the banks of the Rukshan River, at over 900 m.

Iran

In Iran, the altitude limits of the date palm can be determined with some accuracy and with comparative ease, because the roads lead from the shore of the Persian Gulf, where the date palm is very much at home, across the high mountains, whose summits are far too cold for dates, and inland to the big towns.

Along the road from Bushire, 28°59'N, on the coast, inland to Shiraz, there are date palms wherever there is water as far as 7 km beyond Kazarum, 29°37'N, 818 m. In Shiraz 29°36'N, at 1 530 m, there are a few date palms, but they do not mature their fruit.

On the Fasa road from Jahrum to Shiraz, at Fasa 28°57'N, at over 1 200 m, are the last, not very flourishing, date palms, while on the Khafar road from Jahrum to Shiraz, at Khafar itself, 28°57'N, the last dates are at a little below 1 200 m. At Jahrum 28°31'N, at 1 128 m, the date palms grow magnificently; but the dates are late in ripening, and much of the later crop, that harvested in the second half of October, is still in the hard, glossy, yellow stage (khaarak or khalaal) and has to be boiled.

At Darab, 28°46'N, at an elevation of 1 189 m, there is a considerable date grove; but, on the road from there to Shiraz, the date palms continue for only a few km to Hasanabad, which is slightly higher, and then cease.

On the road from Kwash in Baluchistan to Kerman, the date palm stops abruptly at Bam, 29°07'N, at 1 067 m. There they grow vegetatively to perfection and produce some of the finest dates in the world, though most of them are harvested in the soft stage (rutab), as the harvest is late, finishing at the end of October, and only a relatively few dates reach the final, dryer, long-keeping stage (khurmaa or tamar).

On the road from Bandar Abbas on the Persian Gulf to Kerman, the last big grove is at Hajiabad, 28°18'N, at an elevation of 933 m, but the palms do not stop until Aliabad, 28°36'N, at 1 380 m, is reached; here there is a small grove.

The most northerly commercial date cultivation in Iran is at Qasr-i-Shirin, 34°31'N, which lies below 500 m.

The two places, Bam and Jahrus, where the finest Irani dates are grown, dates that rival the celebrated Daglat Nuur of North Africa, are both over 1 000 m high. Bafq and Khur whose dates are eaten in the yellow stage, are also over

Berbera, Somalia. The _hagaa_ or _khariif_, the scorching, south-west wind of summer, blows with gale force on the coast of Somalia. Too wide spacing for date palms therefore is inadvisable.

1 000 m, far north and cold in winter.

In the date zone of Iran, which stretches from 28°07'N in the south to 34°31'N in the north, it would seem that the altitude limit for successful date cultivation is at about 1 200 m or less.

Iraq

In Iraq, the country with more date palms than any other, little information can be obtained concerning the altitude limit, because all that part of the country which is not too cold for dates in winter, is at an elevation of less than 400 m. The only exceptions are the few groves in the northeast where, following the southwest flank of the Kurdish Mountains from Khanaqin, one arrives at the date grove of Tuz Khurmatli, 34°53'N, while the last, few straggling palms are not reached till Kirkuk, 35°27'N. The altitude of both places is less than 500 m.

The Arabian Peninsula

In the Arabian Peninsula higher elevations for date cultivation are common. In Hasa the groves are low, but in Oman most of the mountain oases lie between 500 and 1 000 m. South of the Wadi Hadhramaut at Madi, 55 km northeast of Makalla, at an elevation of 625 m, there is a mixed grove of dates and coconuts, but in the Hadramaut valley itself, where the altitude is between 700 and 800 m, only dates are found. At Sif in the Wadi Duan, 118 km northwest of Makalla, at 935 m, there is a magnificent oasis, with date palms, and some of the mountain oases to the west are probably even higher. There are also groves at Teima, 351 km north-northeast of Medina, probably at about 1 300 m.

North Africa

In North Africa, maximum altitudes on the western side of the continent are about the same as those in Iran, but on the eastern side they are lower. The altitude of the banks of the Nile in its date-growing part is much the same as that of the banks of the Tigris and Euphrates in their date-growing parts. In the north of the Arab Republic of Egypt the date palm country is all below 100 m, and in the south of that country below 200 m, while in the Sudan all groves are below 400 m.

In Libya half the date groves lie along the coast only a little above sea level, and the others are all below 500 m, except those at Mizda on the southern border of the Nafusa Mountains, which are at 510 m, those at Gatrun in the extreme south at 512 m, those at Serdeles, 100 km north-northeast of Ghat, at 630 m, and those at Ghat itself on the eastern flank of the Tassili n'Ajjer at 760 m.

In Tunisia the date plantations are all low. The highest is that of Gafsa, 640 km south-southwest of Tunis, at 345 m, to the north of the habitat of the Deglet Noor variety.

In Algeria north of the Sahara, the date palm is hardly to be found and then only at a high altitude, 538 m at El Kantara (north of Biskra), 578 m at Bou Saada (south-southeast of Algiers), 900 m at Tebessa (south of Bone), and

Date Palms and the Tower of the Kutubiiya at Marrakesh, Morocco, north of the Atlas Mountains. The palms are largely seedlings and the dates ripen here only imperfectly.

1 426 m at Aflou (south of Miliana). At the first of these places the dates do not ripen well and at the last hardly at all.

The Sahara is the real date-growing region of Algeria. Here are the famous groves of the Oued Rhir, nearly at sea- level, with Touggourt at their southern end at an altitude of 69 m, the home of the Deglet Noor, but most of the oases further south are above 500 m. At least one is over 1 000 m, that of Djanet, 1 094 m, on the southwest flank of the Tassili n'Ajjer. Here the dates are of good quality. There is a second grove high up in the Hoggar Mountains, also probably above 1 000 m, that of Tin Guellet, 13 km east of Tamanrasset, which lies at 1 365 m.

In Morocco the date groves which are situated on the southeast side of the Atlas Mountains are mostly above 500 m. That of Bou Denib, 425 km east of Marrakech, is at 935 m, and two others at least are at more than 1 000 m, that of Kaar-es-Suk, 1 061 m, and that of Taourirt de Quarzazate, at 1 146 m, or perhaps a little less.

In Tibesti, Republic of Chad, the date palm flourishes at Aozou, 880 m, and at Bardai 960 m. The highest palm groves are at Dazinga and Goubone, both at 1 500 m (Munier, 298).

From the foregoing it will be seen that the date palm will mature its fruit in North Africa up to an altitude of a little over 1 000 m, and presumably produce at least fresh dates in the south at 1 500 m.

South Africa

The two most promising date-growing districts in South Africa, Henkries Fontein, and Kakamas, are both on the Orange River; the former at an altitude of between 300 and 600 m, and the latter between 600 and 900 m.

Australia

The few date plantations of Australia all lie below 300 m. Those in South Australia (Coward Springs, Hergott Springs, Oodanatta) are below 150 m.

U.S.A

The greater part of the date groves of the United States of America lie between Indio in the north at 6 m below sea level and Mecca in the south at 60 m below sea level with Coachella in between at 22 m below sea level. All these places are just north of the Salton Sea in California. Phoenix, the centre for most of the Arizona groves, has an altitude of 335 m. This is about the limit of commercial cultivation.

Mexico

In Mexico the highest date grove is that of S. Ignacio, at an altitude of over 900 m.

Spain

The Spanish groves are on the east coast a little above sea level.

Israel

The date palm of the banks of the Dead Sea grows and ripens its fruit at 392 m below sea level.

CHAPTER II

CLIMATIC FACTORS

A. GENERAL

The number of plant species increases towards the equator, and decreases towards the poles. Similarly, as one approaches the sub-tropical deserts, the number of plant species rapidly become less. Of those that remain in the centre of the Old World sub-tropical deserts of the northern hemisphere and which survive their rigorous climate, the most striking is the date palm. This palm, unknown in the wild state, although typical of these deserts and widely spread over them, covers only an extremely small part, and produces satisfactory fruit on an even smaller one.

The date palm is not easy to place in one of the usually accepted classes of plants, grouped according to their normal habitat. Thus, this palm, although frequently found growing in sand, can hardly be classed as arenaceous, nor aquatic, although its root is characterized by air spaces like those of the banana root, and it grows well in situations where the subsoil water is very close to the surface. Nor is it a true halophyte, despite its healthy growth in exceedingly salty places, as it grows better where the soil and water are sweet. Nor is the date a xerophyte, since, although it has a thick, waxy cuticle, leaflets with the two halves of their upper surfaces to the sun, reduced leaf surface, and remarkably efficient insulation of the growing point and of the vascular bundles of the trunk, yet it needs a copious water supply. It has not the tiny, dry head above ground and the immensely long tap-root of a typical xerophyte, such as *Caligonum Camosum* or *Pieris Sahariae*. Its roots are adventitious and, though they grow horizontally a long way, the palm suffers when they have to search for water at a depth of more than about two metres.

The five most important climatic factors which determine the suitability of a site for horticultural purposes, are temperature, rain, atmosphere, humidity, light and wind. The effect of these factors on the growth of the date palm will be discussed in the following paragraphs.

B. THE EFFECT OF TEMPERATURE

1. Average Temperatures

The average daily temperature is usually given in climatic tables as the average of the daily maximum and daily minimum. The average of a larger number of records for each day would give a more accurate result, but the average of the two extremes is probably near enough to the true average to make that figure sufficiently useful for comparing the climates of different places with regard to their suitability for date cultivation. The annual average of the daily averages then give a useful filter, as it were, for the separation of those places which might grow dates, from those that are probably unsuitable.

Table 3 gives the annual average temperatures for various date growing regions.

Table 3

AVERAGE ANNUAL TEMPERATURES FOR DATES IN VARIOUS PLACES

Place	Country	Authority*	Temperature °C	Period
Basra	Iraq	1	24.2	1937-1941
Baghdad	Iraq	2	22.5	1937-1946
Tozeur	Tunisia	3	21.3	1901-1950
Touggourt	Algeria	4	21.45	1913-1938
Biskra	Algeria	4	21.8	1913-1938
Niamey	Republic of Niger	5	31.5	?
El Facher	Republic of Mali	5	24.8	?
Kidal	Republic of Mali	5	27.7	?

* 1 - Iraq, Prin. Bur. Stat. (1941)
2 - Iraq, Prin. Bur. Stat. (1946)
3 - Tunis, Service Tun. Stat. (1953 ?)
4 - Seltzer (401)
5 - Munier (299)

2. Minimum Temperatures

Mason (271) states that mature date palms at San Antonio, Texas, survived a temperature of 4°F (-16°C) for a short while, although the foliage was killed, that many hundreds of date palms in the Coachella Valley, California, survived minimum temperatures between 13 and 15°F (-11 and -9°C) on two days in 1913, and that growth was not checked by a short spell at 21°F (-6°). In the Coachella Valley, in 1937, between 3 January and 10 February there were 26 days on which the temperatures fell below freezing, and the total time that the temperature remained below freezing was 14 hours. For 4 hours and 50 minutes the temperature was below 20°F (-7°C), and the lowest temperature recorded was 13.4°F (-11°C). Severe damage was reported to most of the Valley's 146 000 date palms (Nixon, 319). Nixon further stated that he had never observed damage to dates when the minimum temperature was above 20°F (-7°C). He also noted that gardens that were irrigated during the freeze were less damaged than the dry ones.

The Zahidi variety, which is the most northerly of the common Iraqi varieties, was one of the varieties least damaged, according to the speakers at the date growers meeting in Coachella at which the freeze was discussed, and the Khalasa was mentioned as among the varieties most damaged. The home of the last is much further south than that of the Zahidi, and thus some confirmation is given to the general belief that the distribution of date varieties in the date-growing districts of the Old World corresponds, at least to some extent, to their climatic requirements. See also Nixon (323) where he estimates the intensity of the damage caused by the freezes of January 1949 and January 1959 to 43 varieties growing in the Coachella Valley.

Mason's careful experiments show (271) that the growth of the date palm does not cease if:

(a) the minimum daily temperature does not fall below freezing; and

(b) the maximum daily temperature at the growth centre does not fall below a point between 48°F (9°C), and 50°F (10°C).

For short periods, the air temperature may be below this point, but the time lag between the air falling so low and the temperature of the growing point falling so low may be greater than the period during which the air temperature is so low and, therefore, growth would not stop. As Mason remarks, the sero temperature for growth is rarely reached in date-growing regions, and then only for very short periods.

Official meteorological records for January 1964 for Abadan, close to the head of the Persian Gulf, and for Ahwaz and Kut Abdullah, close together and both about 100 km further away from the Gulf, all date-growing districts are as follows:

Table 4

JANUARY TEMPERATURES, KHUZISTAN, IRAN.

Particulars	Abadan	Kut Abdullah	Ahwaz
Latitude N	30°22'	31°13'	31°20'
Longitude E	48°13'	48°40'	48°40'
Elevation, m	3	15	20
Temperature, C, at 135 cm above ground, monthly average	7.8	7.5	-
Temp. C, at 135 cm, above ground, average of daily maxima	14.5	14.1	-
Temp. C, at 135 cm, above ground, highest	20.2	19.3	20.2
Temp. C, at 135 cm, above ground, average of daily minima	-	-0.9	1.1
Temp. C, at 135 cm, above ground, lowest	-4.5	-8.5	-7.0
Temp. C, at 135 cm, above ground, number of days minima less than C	-	16	10
Temp. C, at 2 cm above ground, lowest	-	-8.5	-1.3
Temp. C, at 2 cm above ground, monthly average	-	-2.8	-

In official meteorological records, 1964, it is stated that on 18 and 19 January 1964 an extraordinary weather situation occured. An unusually deep depression passed over Iran from the northwest, after the passage of which very cold air was transported from the inner parts of Asia over the whole country.

Meteorological readings have been taken at Abadan Refinery from 1914 to 1961. The cold spell that occured from the 19th to the end of January 1964 is the longest and most severe that has been recorded since 1914. [Presumably the records from 1961 to 1964 were kept at Abadan, but not in the refinery.]

No adult date palms, and only a few shoots, were killed by the cold in the above mentioned places, but the number of fronds killed varied from a few in some (protected) palms, to nearly all in the others (more exposed). Perhaps the average number of fronds killed at Ahwaz and Kut Abdullah was about half the total, and at Abadan between a fifth and a quarter. The yield of dates in the autumn of 1964 was generally considered however to have been somewhat heavier than that of 1963.

3. Maximum Temperatures

Vinson (446) appears to have been the first writer to note that "high minimum temperature seems to be a more important factor in determining the maturity of the (date) fruit than high maximum temperature".

No place on earth seems to be too hot for the date palm; but in places where the heat is intense, the northern Sudan and the southern Fezzan, for example, the dates do not ripen into the usual syrupy and sticky stage, but into a very hard and dry form. It is, however, probably the dryness of the atmosphere rather than the extreme heat that hardens the date. The matter will be mentioned again under humidity.

An annual maximum shade temperature of nearly 120°F (49°C) is usual at Basra (Iraq) close to the river and somewhat higher at Shaiba in the desert. The absolute maximum shade temperature at Basra near the river between 1938 and 1952 was 123°F (50.5°C) in July, and at Shaiba 125°F (51.7°C). At Baghdad, 121°F (49°C) was recorded on 21 June 1943, and at Haditha on the Euphrates 123°F (51°C) was recorded in 1940.

The maximum recorded temperature during the years 1913 to 1938 at Biskra at the northern edge of the Sahara was 49.2°C (120.6°F) in August (Seltzer, 1946); at Touggourt, 200 km further south, in the same months and during the same period, it was 49.8°C (121.6°F).

[The report, quoted in successive issues of Whitaker's Almanack, that a temperature of 136°F (58°C) was recorded at Azizia in Libya on 13 September 1922, is open to grave doubt, because a study of the Italian original records shows much irregularity in the administration of the meteorological office in that hamlet in that year. Whitaker in 1965 (p. 1007) makes no mention of the Libyan so-called record but gives the same temperature for San Louis, Mexico, on 11 August 1933.]

N.B.: The temperature of objects exposed to the sun may rise to as much as over 40°C above the ambient air temperature on a clear day in still air, but in a moderate breeze the excess temperature may be no more that 8°C (Iraq, Met. Dept., 1919).

The principle date-growing regions of the world, lying on either side of 30°N latitude are sometimes hotter in the summer than many regions within the tropics; the reason for the heat of the date lands is their low humidity, great insulation, and their long days in summer.

4. Diurnal Range

At Bushire, on the shore of the Persian Gulf, where dates grow well, but are generally eaten fresh, the greatest diurnal range of temperature occurs in November and December, and then amounts to no more than 9°C, whereas, at Baghdad, in Iraq, 500 km from the head of the Persian Gulf, the greatest diurnal range amounts to 17°C, and occurs in the months of August and September. Fischer (168) states that in the Algerian Sahara a daily range of 20°C is almost the rule, while a jump of as much as 40°C has been observed in 24 hours with a change of wind. Seltzer (401) gives the average diurnal temperature amplitude for Touggourt (1913 to 1938) as 14.8°C in July.

Cosson (96) seems to have been the first botanist to conclude that the high degree of <u>thermostasy</u> of the interior of the date palm trunk was due to the influence of the ascending sap stream from the roots, although other investigators (see Mason, 271) had already come to the same conclusion with regard to other trees. Mason was the first to study thoroughly the close correspondence between the temperature in the centre of the date-palm trunk and that of the sub-soil water, and to point out the importance of the massive insulation against extremes of air temperature provided by the woody and fibrous parts of the palm. As Mason rightly remarks, the date is the most resistant of all the palms to extremes of heat and cold.

5. Total Heat

It was De Candolle in 1955 (110) who first discovered that the date palm will flower only when the shade temperature rises to 17.8 or 18°C (64°F), will form fruit only above a temperature of 25°C (77°F), and that it needs a total of 5 100°C of heat units, reckoning from 0°C, to ripen dates completely.

Twenty-four years later, in 1879 (97), Cosson estimated the sum of heat units necessary as 6 000°C, while Fischer (168) calculated that the date palm from flowering to fruiting in Cairo received a total of 6 136°C heat units in the 239 days from 26 February to 22 October, and at Biskra of 6 216°C heat units in the 227 days from 24 March to 5 November. The earlier date in each place is that when the temperature first reaches 18°C in the year, and the later date is that of the last day in the year when the temperature reaches that limit. Fischer adds that Hardy, a former director of the Botanical Garden at Hamma near

Algiers, reckoning the period from the beginning of ripening to the complete ripening of the fruit as 214 days, calculated that the palms received 6 362°C of heat units. Swingle (419) made similar calculations in order to estimate whether various places in the southwest of the USA would be hot enough for dates. He took as his length of fruit growth season the 184 days from the beginning of May to the end of October, and used the daily maximum temperatures minus 18°C, which he considered gave a more reliable guide than the temperature reckoning from 0°C, since the flowering process does not commence below 18°C. His calculations for various places in the Sahara and for Baghdad were as follows:

Table 5

HEAT UNITS AND PRODUCTION - SAHARA AND BAGHDAD

(After Swingle, 1904 [419])

Place		Heat Units (C)
Laghouat Algeria	at 792 m on southern slope of outlier of Atlas Mountains north of Sahara. (30 000 date palms in whole district); Too cold for good date production	2 337 (= 4 243°F)
Biskra Algeria	at 122 m, close to southern slope of outlier of Atlas Mountains. 150 000 palms (800 000 in whole district). Excellent dates, but not quite hot enough for best quality Daglat Nuur.	3 049 (= 5 489°F)
Ayata Algeria	40 km N of Touggourt. Less than 69 m. Swingle says Daglat Nuur matures well if summers are hot but ripens imperfectly during cool years.	3 295 (= 5 932°F)
Touggourt Algeria	69 m, 170 000 date palms, (1 690 000 in whole district). Excellent Daglat Nuur.	3 666 (= 6 599°F)
El Golea Algeria	383 m, 7 000 date palms, (80 000 in whole district). No (or few) Daglat Nuur.	3 990 (= 7 182°F)
Baghdad Iraq	34 m, 2 211 000 date palms in whole district (Hussain, A.A.) Excellent climate for first-class dates.	3 898 (= 7 017°F)

Cook (93) in the Coachella Valley, using 1 December as his starting point and continuing for 242 days, that is, to the beginning of the harvest in the Coachella Valley at the end of September, and using the daily mean temperature in degrees F, and subtracting 50°F from each daily mean when it was above that figure, puts the total heat units at 6 900°F. Cook subtracts 50°F instead of Swingles 64.4°F (18°C), De Candolle's limit below which the palm does not set fruit, because for his daily temperature he uses the mean (instead of Swingles maximum) which, he considers, is on the average of about 14.4°F below the maximum.

His figures, therefore, are moderately comparable with those of Swingle, if one consideres the heat units in October (which Swingle includes but which Cook does not), to be about as many as those of December to April inclusive (which Swingle does not include but which Cook does).

Munier (299) considers the fruiting period to last for 180 days for the so-called common varieties: Gharas, Daglat Beida, etc., and 200 days for the Daglat Nuur. He gives as the "norme denommée chaleur fructification", at Touggourt, 1 800°C for the "common" varieties and 1 890°C for Daglat Nuur.

Swingle made the total heat units for the dates in that place 3 666°C. These two figures agree pretty well considering that Munier uses the mean daily temperature and had access to a much shorter period of records (Seltzer 401).

For the present purpose, which is to correlate temperature with fruit production, one might expect the mean daily temperature to be better datum than the maximum temperature from which to calculate the amount of heat to which the palm is subjected.

Table 6 has been compiled mostly from Munier (298). It gives his calculations of the heat units received by dates from pollination to harvest at several, widely separated places where dates are grown in Africa north of the Niger.

Table 6

METEOROLOGICAL AND OTHER DATA FOR VARIOUS PLACES WHERE DATES ARE GROWN NORTH OF THE NIGER

Place Country	Touggourt (Algeria)	Atar (Mauritania)	Bilma (Niger)	Kidal (Mali)	Nema (Mauritania)	Kayes (Mali)
N. Latitude	33°9'	20°38'	18°50'	18°27'	16°50'	14°26'
Longitude	6°2'E	13°12'W	13°30'E	1°21'E	7°00'W	11°28'W
Elevation m	69	229	305	479	266	below 200
Aridity Index	1.95	1.34	?	?	5.33	25.45
Heat units	1 800–1 890	1 875	1 950	1 980	1 903	1 980
No. of days, pollination to harvest	180–200	150	150	135	135	150
Date of pollination	?	1 March	15 March	1 March	15 Feb.	1 Jan.
Date of harvest	?	31 July	15 August	15 July	30 June	31 May
Date Rains begin (over 10 mm a month)	Nov.	?	?	May	?	?

C. RAIN

Rain often does good by washing away some of the salt in the soil, which is the chief trouble in much of the date country, but if it occurs during the period, a few hours before, or a few hours after pollination it may seriously reduce the set of the fruit, unless the pollination is repeated when the weather dries up. (See Nixon, 333 Chapter 10). The chief damage caused by rain occurs when either the rain is early or the dates are late in ripening. Rain does not always damage the dates when they are in the hard glossy, green or yellow (and red) stages, although it can sometimes cause severe checking, and cracking in the late green (Kimri), and coloured (Khalal) stages. When the dates are in the soft "rutab" stage however, rain and the associated humidity causes severe damage including rotting, and fall.

In almost all places where dates grow, the winter rains start sometime before the last of the dates are harvested. The extremely late varieties like Khasaab and Hilaali, however are not damaged, because they are still in the hard glossy stage and will be eaten like that. Some early varieties, for example the very early Al Mehtari of Minab, on the southern coast of Iran which, like the dates in Kidal, Mali (see the section "Total Heat"), take only 135 days from pollination to ripening, entirely escape rain damage. In other cases, where the rains are early, much of the crop can be saved by being cut in the hard yellow (or red) stage and boiled. Drying on the ground after boiling takes a week or 10 days. The dates are then sacked and stored under shelter. Alternatively, crops harvested before being properly cured on the palm may be finished off on the roofs or in the yards of the cultivators, being taken inside their huts when rain is imminent, and put back outside when the rain stops.

At Minab and in Makran in Iran, much of the crop of rutab is passed through simple but effective man-power presses to expel the syrup, which is then preserved in earthenware jars, while the now drier dates are stamped into baskets. Both the syrup and the dates can be kept for several months, whereas, if the soft dates were left in that state, there would be much wastage.

Varieties of dates differ in their susceptibility to rain and himidity. The Deiri (Dayri) variety for example, which in Basra is a hard date, in the climate of the Coachella Valley of California is semi-dry, its quality being improved by the higher humidity of its new home. Nixon (325) found it to be one of the least damaged of sixteen varieties by rain and humidity.

The foregoing shows that little can be learned of the suitability of any particular place for date growing by a study merely of annual totals of rainfall. Autumn rains are the important ones for the date grower; spring showers at pollination time can generally be dodged or pollination can be redone.

Table 7

RAINFALL IN VARIOUS DATE GROWING PLACES

Country	Place	Period	Rainfall mm July	Aug.	Sept.	Oct.	Nov	Authority*	Harvest begins
Pakistan	Multan	?	60	50	8	0	2	3	
Oman	Masqat	1903–1940	0	0	0	2	10	1	
Iran	Bushire	1886–1938	0	0	0	2	40	1	
Bahrein	Bahrein	1917–1938	0	0	0	0	10	1	
Iraq	Basra (Shaiba)	1914–1938	0	0	0	2	22	1	Mid-August
Iraq	Baghdad	1888–1918	0	1	0	2	21	2	Mid-September
A.R.E.	Cairo	?	0	0	0	1	3	2	
Tunisia	Tozeur	1901–1950	1	2	7	9	12	6	
Algeria	El Oued	1913–1938	0	0	3	7	13	5	
Algeria	Touggourt	1913–1938	1	1	2	5	11	5	
Algeria	Biskra	1913–1938	1	2	7	9	12	5	
Niger	Niamey	?	133	226	83	24	0	4	
Mali	Kidal	?	37	65	20	3	0	4	Mid-July
Mali	El Facher	?	114	136	30	5	0	4	

* 1 – Hodgson (202)
2 – Iraq Met. Dep. (214)
3 – Milne (277)
4 – Munier (299)
5 – Seltzer (401)
6 – Tunis Ser. Met. (436)

Table 7 shows that the main date-growing regions are almost rainless until November, with the exception of Multan, where the monsoons break in July, and where the dates are mostly eaten fresh, and with the striking exception of the three places in the Republics of Niger and Mali, where there is heavy rain in July, August and September, and where November is dry. Munier does not state whether the dates cure on the palm in these two Republics, but one would suppose that they are harvested at an early stage.

D. HUMIDITY

The date palm is par excellence the fruit tree of the arid lands of the torrid zone. Air humidity has a considerable influence on the date palm. Where it is high the well-known date palm fungus, Graphiola phoenicis (Moug.) Poit., is common and the well-known pest, the Date Mite, is absent. Where it is low, the fungus is absent and the mite common.

Where the humidity is high, the cured dates are soft, rutab stage, which precedes that of the cured date in a more normal environment. Where the humidity is very low, but the heat great, the date cures into almost stone-like hardness.

The influence of humidity is plainly seen at Basra in Iraq, where the prevailing summer wind is that which blows from the northwest over the dry, bare, scorching deserts of Syria, Jordan, Arabia and Iraq. If this wind blows continuously up to the date harvest, the dates are dry and hard. The Hallaawi variety is especially susceptible to damage by excessive dry wind. The dates shrivel at the apices, the bases remain yellowish, and the flesh is hard. /The name given to these dates is Abu'l Khusheim, "The Father of the Little Nose"; with reference to the yellowish shiny shoulder./ On the other hand, when there are long spells of calm days with a slow drift of almost saturated air up from the Persian Gulf, the dates become soft, sticky, difficult to handle, and are easily damaged. The best results from the point of view of grower, packer, and merchant are obtained in those years, which are in a majority, when the northwest wind, the Shamaal, blows steadily in the early part of the ripening season, followed by a shorter period of the south-westerly drift, the Sharji (Sharqi), and with 10 days or so of Shamaal to finish the curing process.

At places such as Bahrein, and Minab in Iran, where the humidity is very high during the whole of the ripening season, great loss is caused by the falling of the soft dates to the ground. High humidity just before harvest can do more damage than showers of rain followed by bright sunshine and drying winds.

The measure of humidity which would appear to represent numerically most simply the relationship between the texture of the ripe date and the humidity of the atmosphere is the average monthly percentage relative humidity for the months of July to October. Relative humidity may be defined as the ratio of the mass of water vapour per unit volume of the air to the mass of water vapour per unit volume of saturated air at the same temperature (Uvarov and Chapman, (441).

A striking series of old records from the Gulf area is provided by those from Baghdad in the north, where the desert influence is (or was, when the records were taken) strong, from Basra which, although close to the sea, is chiefly influenced by the northwest desert wind, and from Bahrein, a small island in the midst of an extremely hot and steamy sea. The records are given in Table 8.

Table 8

RELATIVE HUMIDITY AT 08.00 HOURS, BAGHDAD, BASRA AND BAHREIN.

Iraq, Meteorological Department, 1919

Place	N Lat	Period of Record	Averages of % Relative Humidity			
			July	Aug.	Sept.	Oct.
Baghdad	33°21'	1888-1918	37	40	42	51
Basra	30°25'	1900-1918	51	51	55	60
Bahrein	26°15'	1905-1919	68	74	74	77

Notes: Baghdad — Chief variety, Zaahidi, rather tough. Excellent Khastaawi, Khadraawi (Baghdad type) etc. Hard varieties (chesib), e.g. Ashrasi, common.

Basra — Excellent dates cured on palm just right for packing direct into cartons or cases without any hydration or dehydration.

Bahrein — Few varieties, e.g. Hallau, cure on palm. Most only reach rutab stage on palm and have to be dried in the sun on the ground.

Table 8 could have been extended to include the Sultanate of Muscat where the average daily relative humidity at 08.00 hours is 81, and which is well known as a date producing country, but to have done so would have been misleading, because the meteorological station is in the town, close to the shore of the bay, and surrounded by mountains. The air of Muscat town in summer, therefore, is rather like that of a minute kitchenette on whose stove a large cauldron of water is boiling vigorously. Where the dates grow, however, is along the 180 km shore to the northwest of Muscat. The gardens are bounded on the west by copen desert, from which come the prevailing winds, more especially of course, the night wind, so that the dates ripen reasonably well on the palms. For the best quality, however, one has to go inland to the mountains of the Fardh country, less influenced by the daily off-sea wind.

The temperature and humidity within a date grove are different from those in the encircling desert and the bigger the grove, the bigger, up to a point, are the differences. Returning in the evening from the still hot, dry desert to the date gardens, one notices their comparative coolness and dampness, especially if much alfalfa is growing below the palms. Meteorological records taken outside the gorves, therefore, even in places only a little way into the desert, for example Shaiba aerodrome near Basra, do not exactly represent the conditions within the groves.

E. LIGHT

Mason (272) has shown that light rays at the violet end of the spectrum inhibit the growth of the date palm which, therefore, grows chiefly by night, and that it is the rays at the other end of the spectrum that are most active in promoting photosynthesis.

The intensity of sunlight may be reduced by clouds. On the whole, the sky is un-clouded in the date countries during the ripening period. Lasserre (241) gives the average number of cloudless days from July to October during the years 1916 - 1920, in three of the most famous date districts of the northern Sahara, as Touggourt 38, Biskra 52, and El-Oued 75.

F. WIND

Wind does not seem to trouble the date palm much. In a gale, whereas bananas growing below the palms will have their leaves torn to shred, the date palm shows no damage. It may be, however, that the nearly black, indurated patches sometimes seen on the developing green dates are due to wind beating the tender dates in their earliest stage against the hard fronds, and not merely to rough handling of the female inflorescence at pollination.

Wind as a carrier of dust and sand is objectionable. Such dust is especially noticeable when it lodges in the webs spun on the date bunches by the Date Mite. Sand and dust do not adhere to the date in its glossy, green, and yellow stages, but they can, and do, adhere to the dates in their soft stage, though much less than one might expect. It has been suggested that mites are carried from palm to palm by wind.

A strong wind may occasionally blow down and old palm if it is very tall, has a large crown, and is growing in shallow soil. A palm can also be blown down if a large number of off-shoots have been removed from all round its trunk at one time, thus depriving it suddenly of its support, or if the Bandicoot Rat has gnawed away the roots on one side of the palm.

Wind can reduce the temperature of the parts of the palm exposed to the sun by convection and by increasing evaporation at the stomata.

Strong, hot, dusty summer winds from the desert can dry up vegetation. Young tomatoes and other vegetables can be killed and citrus and other fruit trees damaged. The date palm however, withstands such winds with equanimity, and is particularly useful in the oasis in breaking the force of the wind, and sheltering more susceptible vegetation. (Langronier and Monciero, 240).

Wind creates sand dunes, and causes them to travel and bury date gardens and the villages dependent upon them. Sometimes the dunes will not entirely cover the palms, whose crowns, once perhaps 10 m above the ground, appear only a short way above the sand. Sometimes the dune may travel on and the trunks of the palms will gradually emerge. If the sand has covered the trunks for a long time, a mass of roots may be found hanging from them from ground level to near the crown.

The chief labour of date cultivators in the Oued Souf, to the east of Touggourt in the northern Sahara, is the daily and continuous carrying up of sand from the great pits in the dunes in which the palms grow and into which it has been blown by the wind, and emptying their baskets and donkey panniers over the crest of the encircling walls of sand.

CHAPTER III

THE ECONOMIC IMPORTANCE OF DATE PRODUCTION

A. WORLD PRODUCTION AND TRADE

Reference to FAO records (see Table 9) indicates that during the last 18 years, world production of dates has oscillated between about two million metric tons and two and a half million tons. There has been a general, although irregular, upward trend as exemplified by the average of 2 022 000 M.T. for the five years 1961-1965 and 2 409 600 M.T. for the five years 1974-1978.

The table shows that the share of the different countries in this total has remained roughly similar over the period. Egypt, with around 400 000 to 460 000 tons has remained easily the largest producer in Africa, followed by Algeria (between 120 000 and 200 000 tons), Morocco (50 000 to 100 000 tons) and the Sudan (down from about 160 000 tons to 110 000 tons).

In Asia, the largest producers are Iraq (up from about 330 000 to 580 000 tons) and Iran (about 300 000 tons), followed by Saudi Arabia (240 000 to 300 000 tons) and Pakistan (up from 90 000 to 210 000 tons).

No country outside the North African and Near East region produces anything like these quantities. In fact the only producer country of any importance outside the region is the U.S.A., with 20 000 to 23 000 tons per annum.

The maintenance of levels of production in the chief producing countries and the increase in production in most of them, might be thought surprising, considering the great difficulties that have been suffered by the industry during the last 20 years.

In many countries of the North African and Near East region, the enormous development of the oil industry has resulted in a flight of labour from the date gardens and a general movement of population from rural areas to the towns. In some of the date-producing areas, it has proved difficult to obtain either regular labour or seasonal labour for pollination and harvesting and in most cases, wages have increased remarkably.

New irrigation schemes have also caused great changes in some countries. For example in Egypt and the Sudan, the Aswan and High Dam caused the flooding of many thousands of palms and the increase in irrigation systems in central and northern Iraq seems to have damaged the production of dates along the Shatt-el-Arab. Pests and diseases have also taken their toll and a comparatively new factor has been the spread of the lethal Bayoud disease in the Maghreb (North West Africa). In the "Sahel" region of West Africa, dates have suffered along with other crops from the drought years.

Table 9

WORLD PRODUCTION OF DATES 1961-78

The following figures are extracted from the 1976-1978 FAO Production Yearbooks giving estimates of annual production for nearly all date-growing countries of the world (1 000 M.T.)

Country of Region	1961-65	1974	1975	1976	1977	1978
WORLD	2 022	2 194	2 345	2 283	2 563	2 663
Africa	851	870	910	905	946	1 021
Algeria	122	168	182	137	140	196
Benin	1	1	1	1	1	1
Cap Verde	1	1	1	1	1	1
Chad	21	25	25	25	25	25
Egypt	407	396	415	417	461	460
Libya	36	62	65	82	83	85
Mauritania	17	13	13	13	14	14
Morocco	79	93	50	70	70	73
Niger	4	5	5	6	6	6
Somalia	5	6	6	6	6	6
South Africa	-	-	-	-	-	-
Sudan	58	100	102	105	106	109
Tunisia			45	42	33	45
North and Central America	28	25	26	23	27	22
Mexico	7	3	3	3	4	2
U.S.A.	21	21	22	20	23	20
South America (Peru only)	1	1	1	1	1	1
Asia	1 127	1 281	1 391	1 338	1 575	1 605
Bahrain	15	16	16	30	38	38
Iran	305	310	290	300	300	300
Iraq	336	360	496	372	578	581
Israel	1	3	4	4	4	4
Kuwait	-	1	1	1	1	1
Oman	43	55	50	50	50	50
Pakistan	92	175	166	202	210	210
Saudi Arabia	238	260	262	257	265	300
United Arab Emir.	(8)	(11)	(11)	11	12	12
Yemen Arab Republic	60	60	65	70	76	67
Yemen P.D.R.	37	41	41	41	41	42
Europe	15	17	17	16	14	14
Italy	1	1	1	1	1	1
Spain	14	16	16	15	13	13

The demographic changes have naturally affected the market for dates - the new urban dwellers have tended to rely more on grains (rice and wheat) as staple foods and less on dates, but nevertheless it seems that consumption of dates in the region has been maintained and often increased. Effron (141) remarked in 1975 that the export trade had picked up from its previous difficulties and that the chief exporting countries were able to dispose of all their crop - the chief buyers being China, U.S.S.R., India and U.S.A.

Reference to Table 10 however, will show that world trade in dates has shown a decline in 1977 and 1978 although fairly steady for the previous ten years, with total exports varying between about 347 000 M.T. in 1961/65, 324 000 metric tons in 1974 and about 352 000 tons in 1976, but dropping to 274 000 in 1977 and 270 000 M.T. in 1978.

In Africa there has been a decline in shipments from the main exporter, Algeria, from about 24 000 M.T. during 1961/65 to 10 000 to 13 000 tons in 1975 to 1978; the next largest, Tunisia, has increased despatches from about 3 000 tons in 1961/65 to 4 700 - 5 200 M.T. in 1975/1978.

In Asia, Iraq has maintained easily the largest export figure with about 263 000 M.T. in 1961/65 and 242 000 tons in 1974, but dropping to 170 000 M.T. in 1978.

The next largest exporter is Iran with about 25 000 tons during 1961/65 and 32 000 tons in 1974 increasing to 45 000 tons in 1978. Saudi Arabia and Oman have remained constant over the last 3 years with exports respectively of 7 000 to 10 000 tons per annum and 5 000 tons per annum.

Israel has shown a steady increase in date exports from 475 tons in 1974 to 900 tons in 1978.

Effron (141) pointed out at the 1975 Date Conference that Iraq and Iran have remained the major exporters to the world market "first, because they have always produced the largest tonnage; secondly, because the Hallawi and Sair varieties lend themselves to transport over great distances and to storage for long periods with minimum risk of deterioration".

He continues "of course the fact that these varieties grow along the Shatt-el-Arab River, convenient to the ports of Basrah and Khorranshahr, stands them in good stead".

In America, the small date production industry in Southern California is of interest because of its high productivity, advanced mechanization and high returns. Commercially, the situation of the date trade in the U.S.A. is exceptional because although quite a high proportion of the production is exported (between 4 000 and 9 000 tons 1974-1978), the country also imports between 6 000 and 21 000 tons annually, mainly from Iraq and Iran. As Effron (141) pointed out, the U.S.A. imports dates in comparable quantities to its home production, even although it seeks export markets from its home-grown product.

The explanation lies in the fact that the Californian product is a high-priced luxury date - mostly Deglet Noor, sold in a "fancy-pack" - whereas the Iraq and Iran dates are a much lower priced product. /It would not be fair to say that the latter are "inferior"; it is rather that the dates from the Gulf area are a quite different fruit, with a different market./

B. NOTES ON PRODUCTION IN THE MAIN DATE GROWING COUNTRIES

Production statistics given by delegates to the 3rd Technical Conference on the Improvement of Date Production, Processing and Marketing (1975) or extracted from FAO records (see Table 9), are as follows:

Algeria

About 7.5 million palms including more than 900 000 Deglet Noor. Production is estimated at 164 000 tons (FAO 1974-1978 average). Main commercial varieties - Deglet Noor and Chars (Rhars). Exports of 10 000 to 15 000 M.T. are concentrated on Deglet Noor.

Egypt

About 4 million date palms in the Nile Valley and almost 3 million in the oases. This represents a great reduction in recent years partly due to the building of the Aswan Dam and the High Dam. Production is estimated at 450 000 tons of soft semi-dry and dry dates, primarily for local consumption (variety Haiyaani). Attempts are being made to export the Zaghlol variety.

Iran

The number of palms is estimated at 20 to 25 million and annual production at about 300 thousand tons. The majority of date exports of about 30 000 tons annually is the variety "sayir" ("Ista'amraan"), grown in Khuzestan. Main varieties for domestic consumption - Muzaafti, Kabkaab, Shahaani and Mordasang.

Iraq

The number of palms is estimated at 20 million and annual production at 350 000 to 600 000 tons, of which about 240 000 tons are usually exported for human consumption and about 40 000 tons for industrial purposes. Main commercial varieties - Zahdi and Khastaawi in the Baghdad area, and Ista'amraan (Sayir), Hallaawi and Khadraawi in the Basra area.

Libya

According to 1974 statistics, there are 4.6 million date palms of which 2.6 million are bearing. Annual production is estimated at 83 000 tons (FAO 1976-1978 average), all used for domestic consumption. Main commercial varieties - Bikraari in the north, and Taasfirt in the south.

Tunisia

The number of palms is estimated as two-and-a-quarter million, about 33 percent Deglet Noor. Other main variety is Fatiimi (Aliig). Exports between 4 000 and 6000 tons 1974-1978, including an important luxury trade in Deglet Noor. Annual production is about 40 000 tons.

Estimated production in other date-growing countries may be summarized as follows:

Country	No. of Palms	Approximate Annual Production	Main Varieties
Bahrain	1 million	35,000 tons (FAO)	Mirzubaan
Israel	134,000	4,000 tons (FAO)	
Kuwait	39,000	1,000 tons (FAO)	
Mauritania	800,000	14,000 tons (FAO)	Tinterguel
Mexico	150,000	3,000 tons (FAO)	
Oman	1 million	50,000 tons (FAO)	Khubeizi, Khalad Umm Silla
Pakistan	1,700,000	200,000 tons (FAO)	
Yemen Arab Rep.		71,000 tons (FAO)	
P.D.R. Yemen	1 million	41,000 tons (FAO)	Hamraiya
Somalia	200,000	6,000 tons (FAO)	
Syrian Arab Rep.	12,000		
Tanzania	3,000		
U.S.A.	200,000	20,000 tons	Deglet Noor, Khadraawi Zahdi, Barhi

Production figures for the following countries are not available, although estimates of the number of palms (in brackets) have been made at various times: U.S.S.R. (6 000); Peru (2 000); Australia (5 000); Brazil (3 000); Colombia (2 000); South Africa (5 000); Spain (250 000).

Table 10

WORLD EXPORT OF DATES (MAIN PRODUCING COUNTRIES) 1961-78

The following figures are extracted from the FAO Trade Yearbooks for 1976 to 1978

Country or Region	Quantity in Metric Tons						Value : $1,000					
	1961/65	1974	1975	1976	1977	1978	1961/65	1974	1975	1976	1977	1978
WORLD	347,617	323,934	348,920	352,780	274,523	270,485	37,516	68,995	85.453	84,980	82,482	94,302
Africa	32,779	20,540	18,532	17,647	17,516	19,358	9,246	12,408	13,900	14,119	13,770	16,602
Algeria	23,948	15,005	11,468	9,839	9,863	13,000	7,524	8,087	7,562	7,325	6,892	9,800
Chad	300	3	-	-			41	1	-	-	-	-
Egypt	137	352	675	62	58	172	49	207	483	87	31	146
Kenya	9	-	-	265	314		1	-	-	92	131	
Mali	321	57	-	23			42	12	-	4		
Mauritania	3,500	1,000	1,200	1,200	1,300	1,500	410	180	190	190	230	300
Sudan	1,222	10		492	736		144	5		273	474	
Tunisia	3,283	3,999	5,127	5,748	5,272	4,686	1,026	3,859	5,620	6,147	6,012	6,356
North and Central America	4,520	4,283	6,615	8,752	5,274	5,806	1,755	4,123	5,889	8,552	6,631	7,491
Mexico	4	-	17	-	44	-	1	-	7	-	2	-
U.S.A.	4,516	4,283	6,598	8,752	5,230	5,806	1,754	4,123	5,881	8,552	6,629	7,491
Europe (re-exports only)	11,365	8,095	8,741	8,947	9,134	11,049	6,550	11,008	12,599	14,258	15,239	20,008

TABLE 10 (Cont.)

Country or Region	Quantity in Metric Tons						Value : $1,000					
	1961/65	1974	1975	1976	1977	1978	1961/65	1974	1975	1976	1977	1978
Asia	298,953	291,016	315,032	317,434	242,599	234,271	19,965	41,456	53,066	48,051	46,752	50,199
India	20	10	41	49	-	-	6	6	25	25	-	-
Iran	25,539	32,640	33,272	13,508	25,000	45,000	2,157	8,644	7,086	2,384	4,600	9,000
Iraq	262,727	242,221	254,162	285,795	203,000	170,000	17,240	29,433	38,835	39,923	37,303	33,500
Israel	-	475	637	723	904	900	-	367	500	631	957	950
Jordan	19	538	149	447	472	470	1	35	9	58	72	72
Kuwait	214	458	226	249	85	87	15	111	46	55	30	30
Oman	3,320	1,200	5,000	5,000	5,000	5,000	125	156	676	680	750	750
Pakistan	74	98	149	54	-	214	10	29	39	16	-	101
Saudi Arabia	5,186	8,196	18,741	10,434	6,800	10,000	311	1,571	4,980	3,469	1,896	2,900
Yemen P.D.R.	1,500	25	600	72	-	-	75	4	130	30	-	-

Note: Countries whose exports of dates are mainly re-exports of other countries' produce are not quoted separately, but are included in the totals for continents.

Broom and fan from date-palm leaflets. Tripoli, Libya.

Frond bases (karnaaf) sharpened for hammering into sandy walls of well instead of a lining of brick or stone, neither of which were to be found in the neighbourhood. El Ergh, Cyrenaica, Libya.

Baskets, brooms and trays made from date-palm leaflets. Baghdad, Iraq.

C. THE DATE AS A SUBSISTENCE CROP

Important as the date is in commerce, both in the local North African and Near Eastern markets, and in international trade, especially that supplying Europe, U.S.A., China, U.S.S.R. and India, its importance as a subsistence crop remains paramount in many poor communities in desert areas. These areas and the neighbouring "Savannah" and "Sahel" regions are susceptible to periodic famines and many districts are frequently short of food.

In the date-growing districts of the Old World remote from oil operations and large towns, the date palm is still almost the universal provider, not only supplying food, but a list of domestic requirements which Dowson lists alphabetically as "alcohol, bags, baskets, bedding for stock, beds, bird cages and traps, blankets, boats, buoyancy blocks for boats, calf muzzles, cattle feed, ceilings, chaff and straw nets, chairs, chicken coops, cradles, cushion stuffing, doors, draughtsmen, drink, fans, fences, firewood, fishing rods, fishnets and floats, fish traps, flumes, fruit crates, grain stores, hats, ladders, lifebelts, material for upholstering furniture, mats, measuring rods, pack-saddling stuffing, panniers, revetment, road metal in swamps, rope, sandals, scent (such as Tarah Water), sieves, shade for citrus, fruit and vegetables etc., shelters, steps down to water, stools, string, supports for earthen stoves, sweetening, toothpicks, tables, vinegar, water-hoists, well-linings and window frames".

It is clear that the date palm will remain an essential crop for many people in the North African and Near Eastern regions for the foreseeable future.

CHAPTER IV

DESCRIPTION OF THE DATE PALM

A. GENERAL

The date palm belongs to the Monocotyledones of which the Palmae form a family of about 200 genera and about 1 500 species. One of the genera is Phoenix (Coryphoideae Phoeniceae) of which there are a dozen species, all native to the warmer parts of Africa and Asia and including P. dactylifera. The date palm, however, is found only as a cultivated plant, or in abandoned gardens, or sometimes at water holes in the desert, where it has grown from seed left by travellers who have rested and eaten dates there.

The date palm is a dioecious perennial, the females of which normally begin to bear dates at an age varying from four to twenty years, with an average of perhaps five years from the time of planting of the shoot, and continue bearing until their death. This usually occurs somewhat before 100 years, but life may be prolonged to 150. Generally, however, date palms are cut down before they reach 100 years as their yield, when they approach this age, is usually so small as to make climbing them for pollination and harvesting not worth-while.

The date palm is a feather palm and may, therefore, be immediately distinguished from all the fan palms. There are however two other feather palms similar in appearance - the wild date palm and the Canary palm. The wild date palm (P. sylvestris, Roxb.) is usually shorter than the date palm, and does not bear offshoots. Its fruit is small, dry and considered inedible.

The Canary palm (Phoenix canariensis Chab. ex Hort.) does not bear offshoots, has a much thicker trunk than that of the date, a much denser crown, and much shorter leaflets. Its spines begin right at the base of the midrib, unlike those of the date palm, which do not begin until some way above the fibre. The differentiation between spines and leaflets in all but a few varieties of the date palm is clear: in the Canary palm it is less so.

The date palm, if the crown of fronds be included, may reach a height of over 36 m (Blatter 38). Nixon (325) says over 24 m to the growing point. Mason (273) in Egypt measured many date palms and found the tallest to be just over 27.6 m to the growing point, but few palms attain so great a height, for they generally die or are blown down before reaching 20 m.

B. VEGETIVE ORGANS

i. Root

The date palm, being a monocotyledon, has no tap root. Shortly after the primary root has issued from the seed, secondary roots appear. They form a dense cluster and are approximately of the same diameter throughout their length. There is no secondary thickening. These roots produce short

Well-grown Khasaab palm: trunk 90 cm in diameter. Fureisat, Iran.

lateral roots of the same type.

ii. Trunk

The date palm has a vertical, columnar trunk of the same girth all the way up, which does not increase once a full crown of fronds has developed. It is covered with the bases of the old fronds, which make it rough, but, by the time the palm has become old, the bases have weathered and the trunk has become smoother. A very vigorous Shaahaani palm at Jahrum (Iran) has a girth of 211 centimetres. Most palms are only half this size. The trunk is composed of tough, fibrous vascular bundles cemented together in a matrix of cellular tissue which, near the periphery of the trunk, is much lignified.

iii. Frond

The first leaf of the seedling is lanceolate and entire. The fronds (leaves) are pinnate and are borne in a crown at the top of the trunk, where they arise from buds produced by the apical growing point. From ten to thirty-five are produced in a year, and they have a life of from three to seven years. The number borne by a palm at any one time usually varies from 30 to 140 - Nixon gives the upper limit as 150.

A frond on a vigorous variety attains a maximum length of about six m, but most fronds are about four m long. The base of the frond is a tubular fibrous sheath, which encircles the trunk of the palm. The greatest width that the frond midrib at the top edge of this sheath attains is half a metre, but normally it is only half this. Its width rapidly narrows from the base upwards. At the lowest point at which it can conveniently be cut off for inspection, its girth varies from 11 to 35 cm. The frond midrib or petiole, which is more or less triangular in cross-section, with two lateral angles and one dorsal, is bare of spines for a short distance above the top of the tubular sheath, but thereafter it bears spines on either side. Above, i.e. more distally, the spines are replaced by leaflets. There is an intermediate zone where spine-like leaflets or leaflet-like spines are found. There may be a single terminal leaflet or two leaflets forming a V.

The buds that form the fronds are produced from the heart or growing point in a flattish ascending spiral.

From the fibre at the base of the frond to the base of the spine-leaflets is about 28 percent of the whole: the spine-leaflets occupy about 4 percent: the leaflets occupy about 62 percent: and the terminal leaflet takes up about 6 percent of the whole length of the frond; but the proportions vary with the variety and with other factors.

iv. Fibre

The base of the frond is a sheath encircling the palm. This sheath consists of white connective tissue ramified by vascular bundles. As the frond grows upwards, the connective tissue largely disappears leaving the dried, and now brown, vascular bundles as a band of tough, rough fibre attached to the lateral edges of the lower part of the midribs of the fronds and ensheathing the trunk. Varieties differ in the height to which the fibre grows up the central column of unopened fronds and in the texture of the fibre and also somewhat in colour.

AHWAZ 26.4.64

KHADHRAAWI

SHOWING INFLORESCENCES
EMERGING FROM SPATHES

Left to right: Barhi spathe and female inflorescence; male inflorescence and spathe, Ista'amraan spathe; Khadhraawi spathe (splitting). Ahwaz, Iran.

Varieties differ in their spikelets and in the arrangement of the spikelets on the main axis of the bunch. Above are shown empty bunches of the Beiyuudhi, Bikroari, Burunsi and Taabuuni varieties. Tripoli, Libya. The right hand object in each of the four photos is the shadow of the bunch.

v. Spine

The spines, varying from almost nothing to 24 cm in length and from almost nothing to one cm in thickness, are arranged on the two outer edges of the frond. They vary in number from about ten to about sixty. They may be single, in groups of two, or in groups of three. The arrangement is not the same on both sides of the midrib. There is one variety of male in Iran which, like the Canary palm, has a fanlike cluster of spines on each side of the frond at its base close to the fibre.

vi. Leaflet

The leaflets or pinnae, which are scattered, rigid, entirely lanceolate, folded longitudinally, and attached obliquely with their folded base to the common woody petiole or midrib, range in length from about 15 to about 104 cm and in breadth from one to 6.3 cm. They may be arranged in groups of one, two, three, four or five pinnae. In any one group, except of course, in a "group" of one, the leaflets are generally arranged at different angles to the frond midrib, but corresponding members in different but similar groups have similar angles, with the result that, looking along the frond, one sees the leaflets arranged symmetrically in different planes. The number of leaflets on a frond varies from about 120 to some 240.

C. REPRODUCTIVE ORGANS

i. Spathe

The spathes, that is, the envelopes that enclose the inflorescences until they mature, arise as buds on the growing point of the palm in the axils of the fronds. They push up through the fibrous sheaths that form the frond bases. At first green, they later turn brown. They are tough and leathery and lens-like in cross-section.

When the enclosed inflorescence becomes mature, the spathe splits longitudinally. The male spathes are shorter and wider than the female ones. In length spathes vary from about a quarter of a meter to a metre. The number of spathes borne by a palm in a year varies from none to about twenty-five in females and to even more in males, but the average is perhaps a dozen for females and rather more for males.

ii. Inflorescence

The inflorescence, a compound spike, when the spathe splits and reveals it, is seen to consist of a main stem or peduncle, which continues to grow and which varies in length when it is full-grown from a quarter of a metre to two metres or more, and of a number of spikelets arising near the apex of the peduncle.

The female peduncle is at first erect but with the growth and increasing weight of the dates becomes semi-circular so that the spikelets eventually hang downwards.

The glabrous, undulating spikelets, which vary in length from some 10 cm

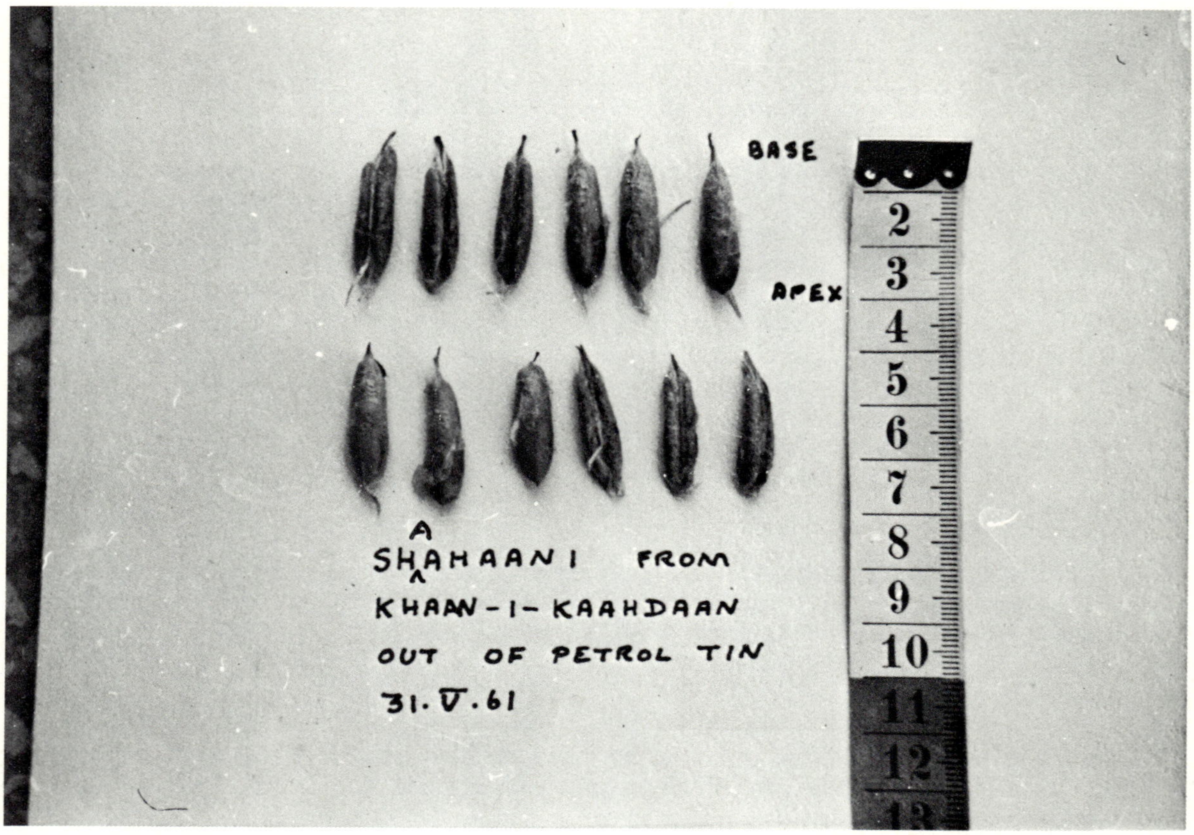

Seeds of the Shaahaani variety of date showing the characteristic "tails". Khaneh Kahdan, Iran.

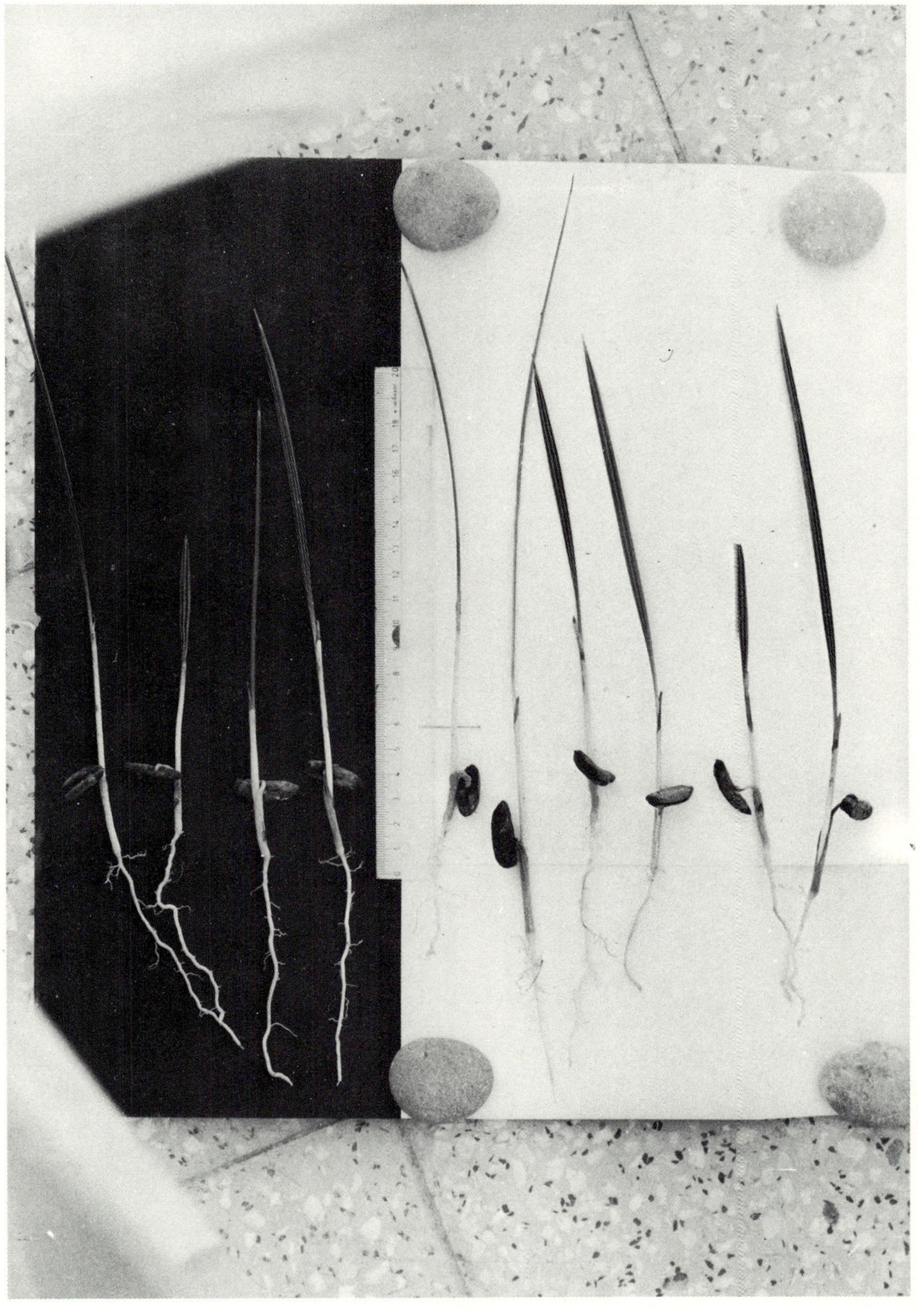

Ista'amraan seedlings. Ahwaz, Iran.

DEVELOPMENT OF DATE-PALM SEEDLINGS

VARIETY OF SEED:-

DAGLAT NUUR, FROM TUNISIA

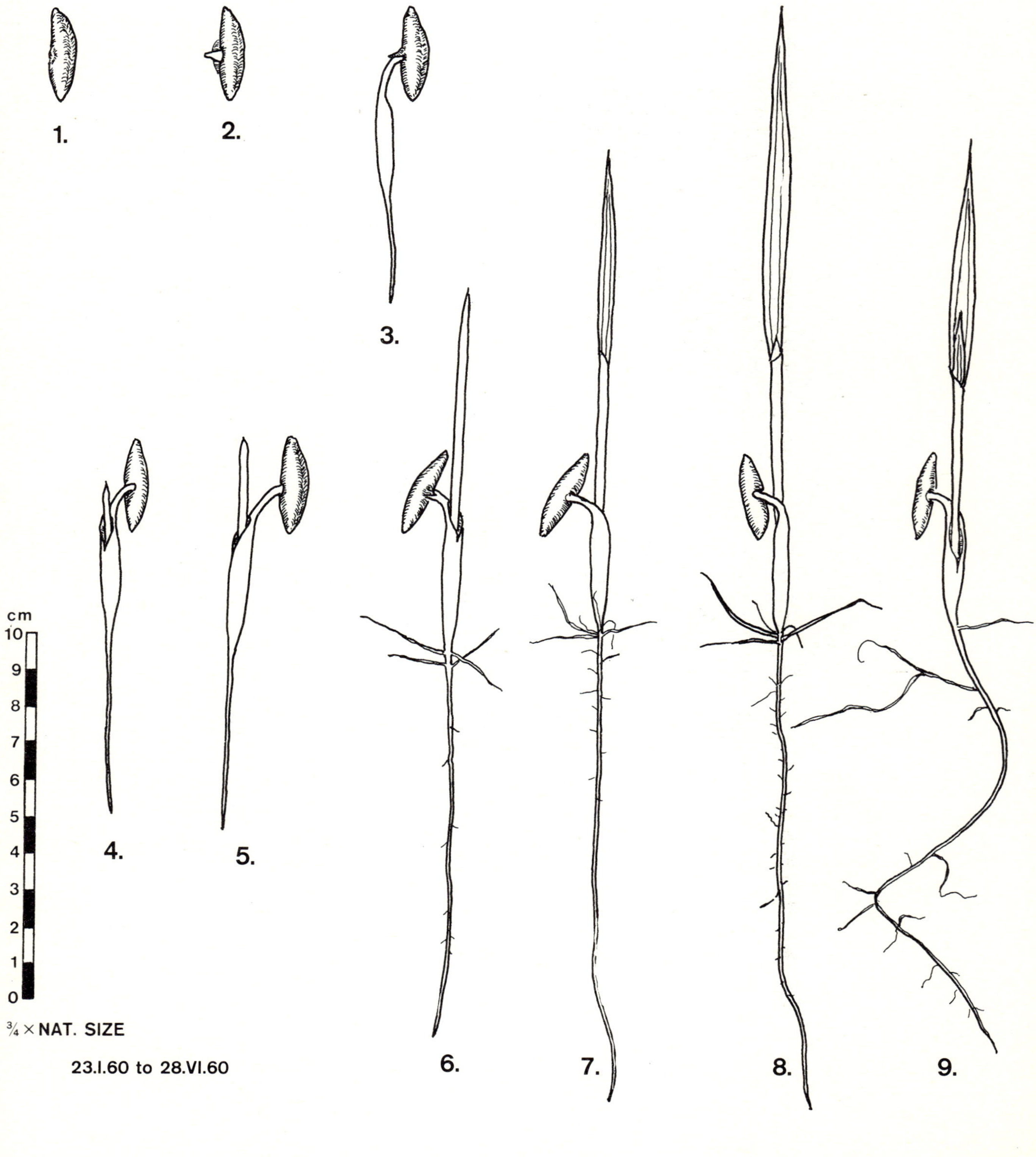

to a metre or more and in number on any one bunch from twenty to thirty to perhaps as many as 150, may all arise at the apex of the peduncle or may arise in groups at intervals between the apex of the peduncle and a point about half way down it according to variety.

The flowers are small, yellowish, coriaceous, sessile and borne at the bends of the spikelets.

The male flowers are sweet scented, have three sepals, connate in a cupular three-toothed calyx; three petals, obliquely ovate and valvate; six stamens with short, subulate filaments and erect, dorsified anthers.

The female flowers have three sepals, connate in a globose, accrescent calyx, and three rounded, imbricate petals, six staminodes, and three free carpels with erect ovules and sessile, hooked stigmas. In pollinated flowers, usually two carpels abort and only one ripens. The aborted carpels persist as two brown spots on the calyx of ripe fruits.

iii. Fruit

The fruit, depending on the variety and the growing conditions, varies in weight from about 2 grams to about 60 grams, in length from about 18 to 110 mm, and in width from about eight to about 32 mm. It is a single, oblong, terete, one-seeded berry, with a terminal stigma, a fleshy pericarp (the flesh of the date) and a membraneous endocarp (the paper-like membrane between the seed and the flesh, which the Arabs of Basra call the thob or "shirt"). In 1941, it was reported at the Coachella Date Growers' Institute that the number of dates on a bunch of 12-year-old Deglet Noor palms in the Arkell Date Garden near Indio in California varied from 288 to 1 406. The weights on the bunches on the same palms varied from 2.3 to 12.7 kg.

iv. Seed

The seed which, according to the variety and the growing conditions varies in weight from less than 0.5 grams to about 4 grams, in length from about 12 mm to 36 mm, and in breadth from about 6 to 13 mm, is oblong, ventrally grooved, with a small embryo, and with a hard endosperm composed not of starch, but of a cellulose deposit on the inside of the cell walls.

The seed of the date palm is longer than that of *P. canariensis*, *P. humilis* Royle, *P. paludosa* Roxb., or *P. sylvestris*.

D. VARIETIES

The number of date varieties is large, perhaps over 3,000. There are about 400 varieties in Iran, a similar number, mostly different, in Iraq, and a rather similar number, all different, in Libya, as well as many additional varieties in the other main date-growing countries.

For the best botanical descriptions of 196 varieties growing in the U.S.A.

SIX COMMERCIAL VARIETIES

of

IRAN

RIPENING STAGES

KEY TO ABBREVIATIONS

H.	= HABBABŌK	= Creamy white
CH. (KH.A)	= CHIMRI (KHALAAL AKHDHAR)	Green, Hard
KH.	= KHALAAL	= Red or Yellow, hard
R.	= RUTAB	= Soft
T.	= TAMAR	= Dry

RIPENING STAGES BARHI

AHWAZ 1965

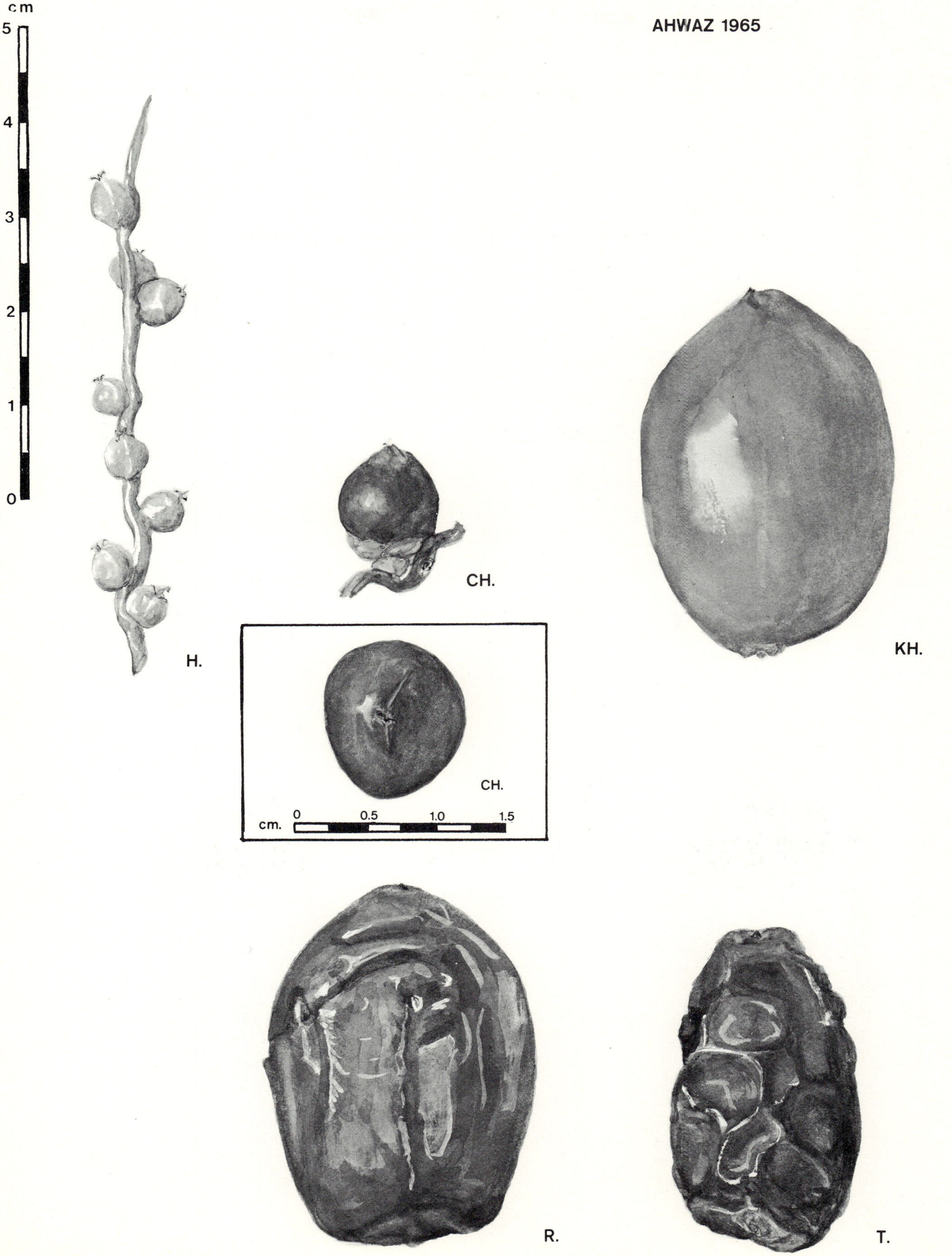

RIPENING STAGES

DAIRI

AHWAZ 1965

CH.(KH.A.)

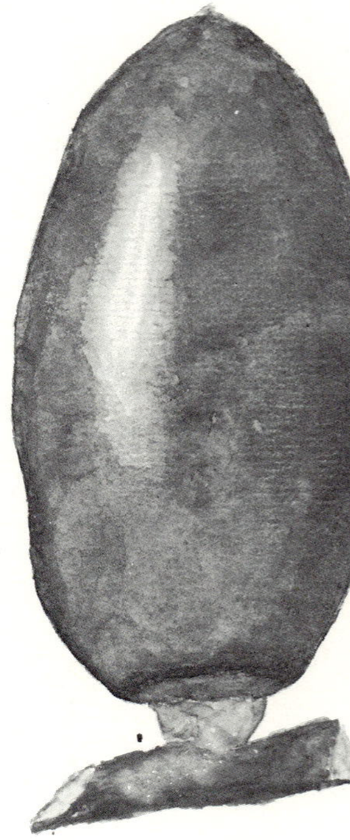
KH.

R.

T.

RIPENING STAGES

HALLAAWI

AHWAZ 1965

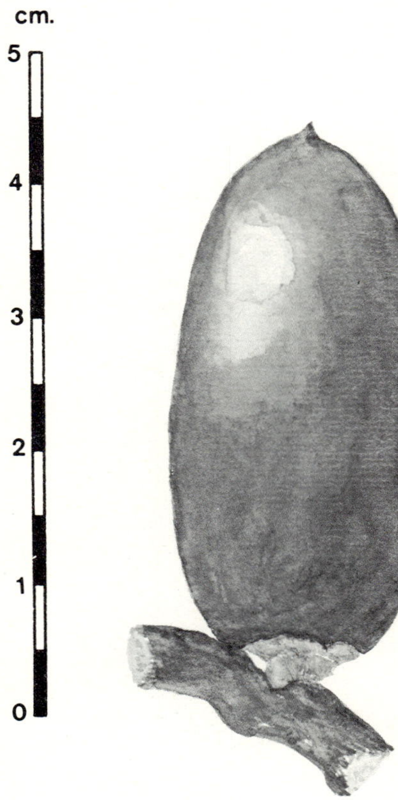

CH. (KH. A.)

KH.

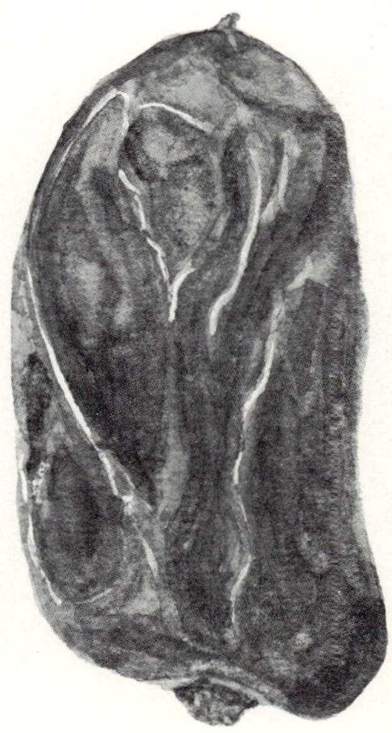

T.

RIPENING STAGES

ISTA'AMRAAN

AHWAZ 1965

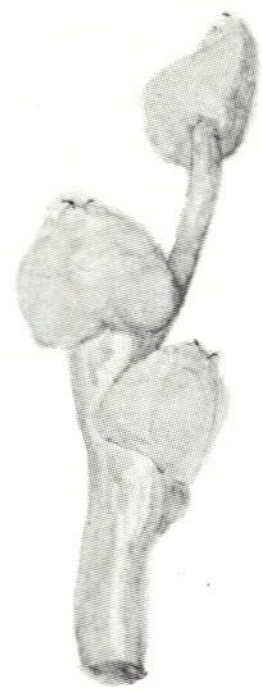

H.

CH.

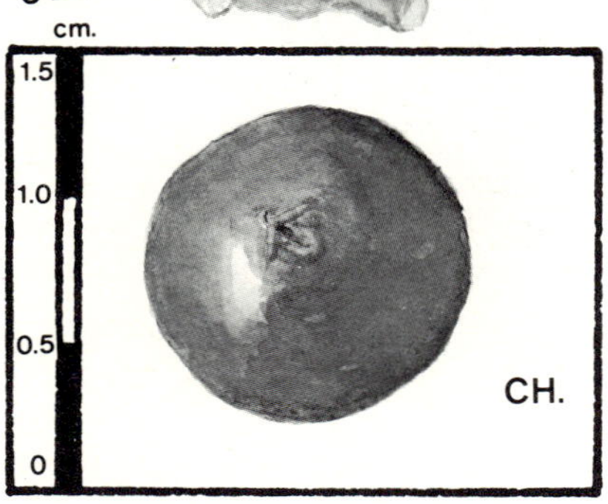

CH.

CH. (KH. A.)

KH.

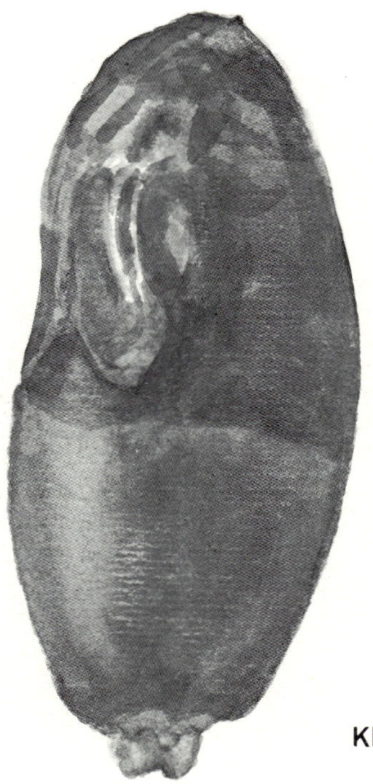

KH./R.

T.

RIPENING STAGES KHADHRAAWI

AHWAZ 1965

CH. (KH. A.)

KH.

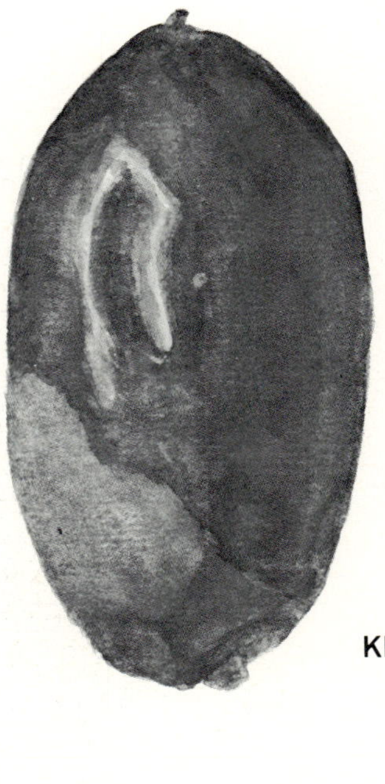

KH./R.

T.

RIPENING STAGES ZAHIDI

AHWAZ 1965

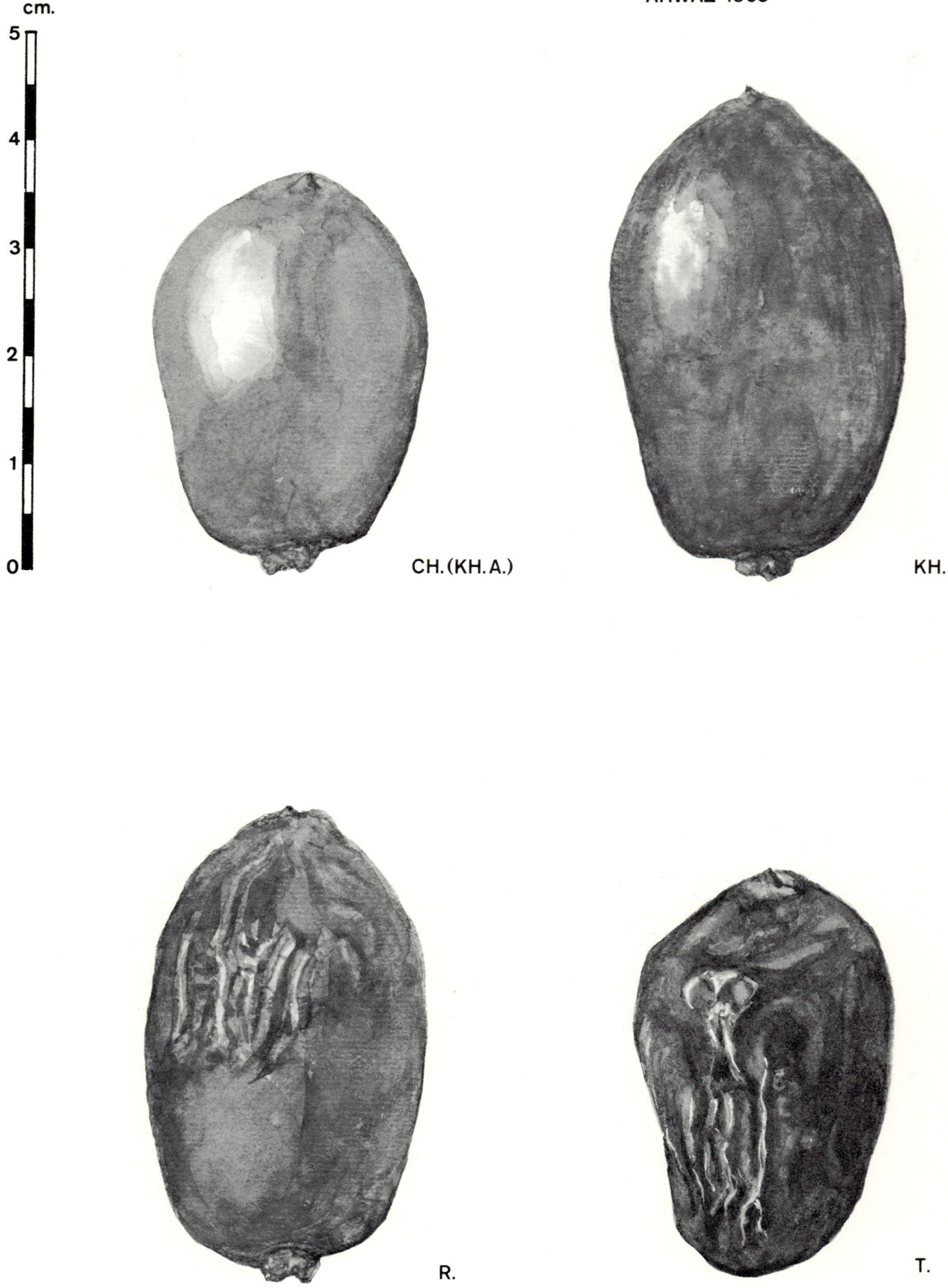

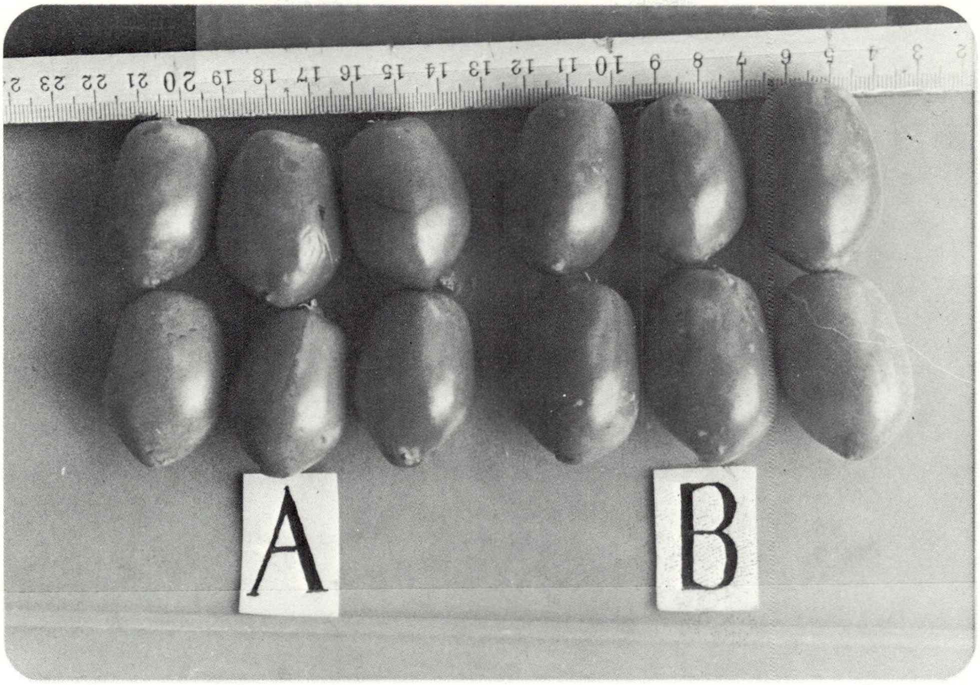

Zahdi (A and B two different stages
turning from Chimri into Khalal stage)

Barhee (Khalal and Rutab stage)

Khastawi (Khalal stage)

Khadrawi Baghdad (Khalal stage and
turning to Rutab stage)

Chipchap (Khalal stage turning to Rutab stage)

Maktoom (Khalal stage turning to Rutab stage)

LIBYAN DATES

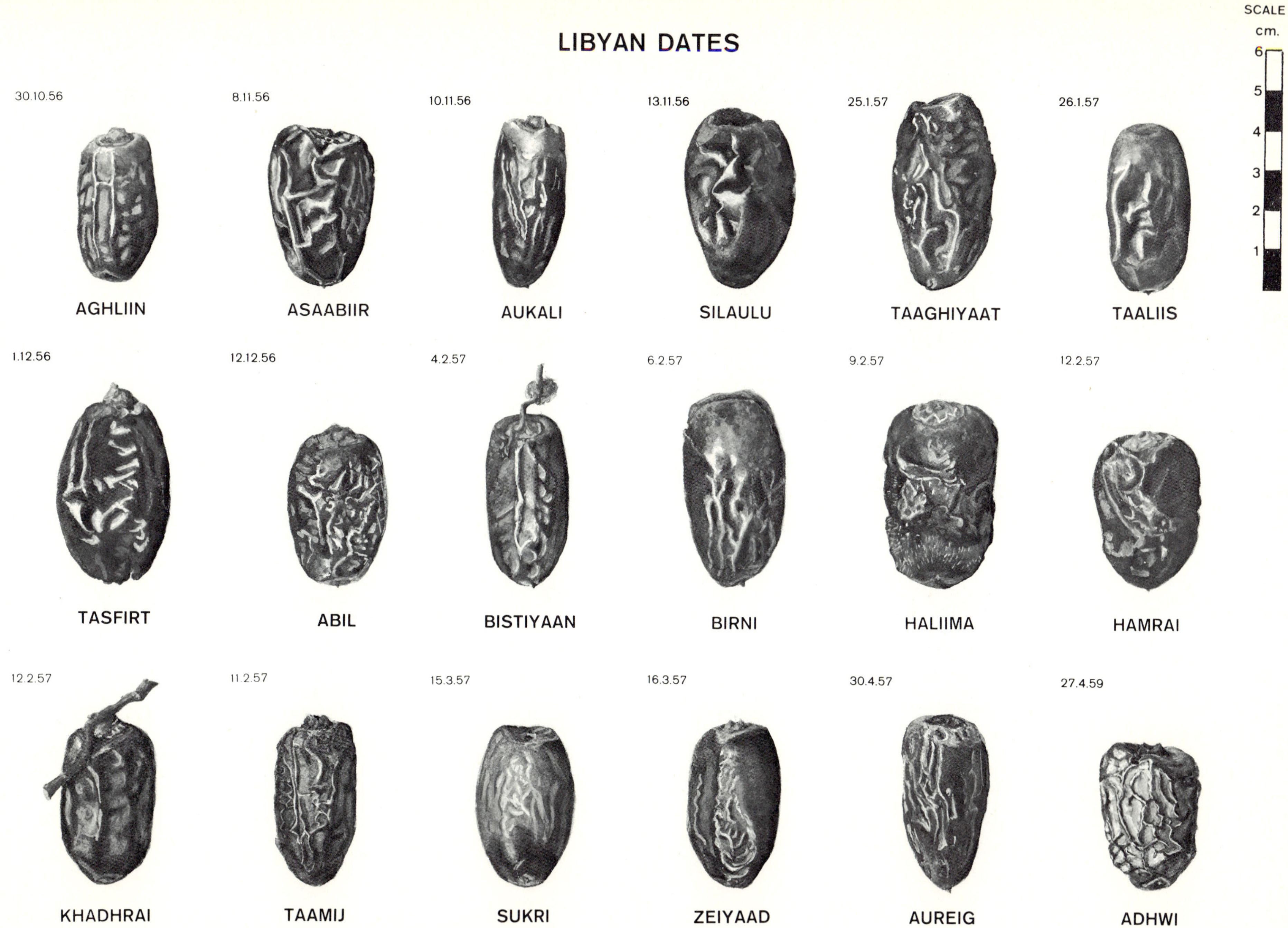

LIBYA 1956 - 1959

18 Varieties of Dates from Libyan Fezzan, Showing Typical

and collected from many places in the world, see Nixon (325). Kearney (226) describes 63 Tunisian varieties. Mason (370) describes 22 varieties of the A.R.E. and the Sudan. For popular descriptions see Dowson (121) for 48 varieties of the Shatt-el-Arab; Brown (61) for 26 varieties of the A.R.E.; and Popenhoe (356) for 100 varieties from different places in the date-growing world, mostly in Iraq, Musqat and Algeria.

Varieties have been developed by thousands of years of selection of seedlings, only those possessing desirable characteristics (early ripening, late ripening, heavy yield, resistance to salt, disease, pests and humidity, etc.) being propagated.

It is interesting to note that at least six of the Iraqi varieties recorded by El-Muqaddasi in the tenth century (985 A.D. was the date he finished his book: his visit to Iraq was earlier) are still, 1 000 later, known there (Khasaawi-el-Baghal, Khastaawi, Seihaani, Sukkar and Tabarzal). Goo (184) mentions that the Nikolsii variety of Jericho has a written history of 800 years and that possibly the excellent variety recently named Jericho is the same as the Nikolsii.

At the head of the Persian Gulf six varieties of males are recognized in addition to the much commoner seedling males (khikri or gheibaani) and at some other places a few male varieties are recognized and propagated because their pollen is considered better than that of the undifferentiated seedling males; but in most places no special varieties of male are recognized.

CHAPTER V

POLLINATION, BREEDING AND SELECTION 1/

A. POLLINATION

1. Antiquity

Artificial pollination of the date palm is a practice that arose very early in the evolution of date culture. The earliest know written records are on cuneiform tablets dating back to the latter part of the third millenium B.C. in ancient Mesopotamia (Scheil, 398). A little later Hammurabi's famous "code" covers it in some detail (Pruessner, 364). Since the code merely gave legal sanction to customs developed over a long period, it is reasonably certain that artificial pollination had been evolved much earlier. The antiquity of the practice in widely separated centres of date culture has led some authorities to believe that it was probably developed spontaneously and independently in several different places (Joret, 221).

In ancient Mesopotamia (Sumer) the date palm was a sacred tree associated with the idea of fertility (Danthine, 104). It is not surprising that in very early times the pollination of the date palm has already acquired a ritualistic importance (Popenoe, 357, Legrain, 248). Even today in some parts of the Arab world a prayer is said before pollination starts (Charlet, 56), and some times special songs are sung (Lethielleux, 254).

2. Methods

Before describing methods of pollination it may be of interest to mention that in at least a few places dates are still produced without artificial pollination. The palms are seedlings and about a half are males, so that there is a fair set of fruit from the natural method of wind pollination. This is the situation at Marrakesh, Morocco, at San Ignacio and nearby communities in Baja California, Mexico, in North America, and at Ica, Peru, in South America. In all three places the fruit is valued and utilized. Some people in these places know about artificial pollination, but there is insufficient economic pressure to bring about change. In Baja California the proportion of male palms in some gardens is reduced to about 15 percent.

A very primitive method of pollination was seen by Belgrave (36) in the oasis of Siwa in the northern Libyan desert. It consisted merely of placing an entire male inflorescence in the crown of the female palm and leaving the result to wind pollination. The same method is sometimes used in Mauritania (Chevalier, 87) and in some parts of Libya (Dowson, 126). It has also been used

---

1/ This chapter was contributed by the late Roy W. Nixon, formerly Horticulturist and collaborator, United States Department of Agriculture, U.S. Date and Citrus Station, Indio, California.

Pruning Fronds to Expose Spąthe for Pollination

Pollinating Female Date-Palms, using Easy-To-Carry Pollinator.

Male date inflorescences drying in the sun. When they are dry they will be shaken over paper and the pollen put in muslin bags which will be tied to sticks and used for pollinating. Shahdad, Kerman, Iran.

sometimes in California for young seedling palms in closely spaced plantings where the objective was primarily to obtain a few specimens of fruit for evaluation. Its use does not necessarily imply ignorance of better methods but may result from scarcity of labour or lack of economic incentive. Obviously this method could not be depended upon for uniformly good sets of fruit unless large amounts of pollen were available throughout the flowering season. However, it could very well have been an intermediate step in the evolution of artificial pollination. Sooner or later someone would notice that female palms near male palms were more fruitful than others, which might have suggested the function of pollen and the idea of placing male flowers nearer the female flowers.

The most common method of pollination is to cut the branches or strands of male flowers from a freshly opened male inflorescence and to place two or three of them, inverted, between the strands of the female flower cluster first shaking some pollen over it. Sometimes the male strands are merely laid across the top of the female cluster, wedged in among the tips of the female strands. When the male strands are placed lengthwise between the female strands they are usually held in place by a strip torn from a leaflet nearby. This leaflet may be tied in such a manner that it breaks or loosens later as the dates begin to grow, or the cultivator may climb the palm two or three weeks later and cut it. In the United States twine is placed around the pollinated cluster and tied in a slipknot with the free end long enough to permit later adjustment to the maximum size of the bunch.

In some places a stick, something only a metre long, called a "laggaha" in Egypt (Brown and Bahgat, 62), is used to facilitate pollination with fresh flowers. In Elche, Spain, the stick is commonly made of a date leaf midrib and has a hole in one end through which is attached a loop about 12 cm in diameter made from a section of a date leaflet. The operator climbs the palm by means of a rope girdle and when he gets up to the lower leaves, rather than make the extra effort of climbing through them, he reaches through the centre of the crown with the stick and puts the loop over the top of a spathe just beginning, or about, to split and pulls downward.

This breaks the spathe and frees some of the strands within. He then takes a sprig of male flowers, puts it in a slit in the other end of the stick and places it in the female flower cluster. If the flowers have already begun to emerge from the spathe the first part of the operation is unnecessary. Pollen is plentiful in Elche and a section of the male inflorescence with 10 or 12 strands is commonly used for pollination (Nixon, 324).

When held for several days before being used, male flowers begin to dry and then shatter more easily. When it is necessary to use dried flowers they are sometimes sprinkled with water or dipped in it (H.W. Brown and Bahgat, 62; Bonavia, 45).

When pollen is scarce the pollinator in Iraq (Dowson, 119) and in Egypt (Brown and Bahgat, 62) sometimes conserves it by placing the flowers, usually dried and crushed, in a muslin bag tied to the end of a stick. The bag is simply shaken over the newly opened female inflorescence and enough pollen sifts through to effect pollination. Water is sometimes sprinkled over the female inflorescence when dry pollen is used.

In California, while a few growers use fresh male flowers for pollination, no one uses dried flowers; most of the growers use dried pollen. This dried pollen is generally applied by dusting it on cotton and placing one or two pieces about the size of a walnut between the strands of the female cluster. A few growers who are well supplied with male palms sift the pollen and apply it with insect dusters, sometimes in addition to that applied on cotton (Nixon, 333).

The use of a rubber bulb blower for pollination was suggested by Bonavia (45). The idea was carried out early in the development of the California industry by one grower who used a copper tube mounted in a bamboo pole and blew dry pollen through it with a rubber bulb from a bettle below (Reed,). A few individuals have continued to use this method with some modification. In one variation the blowing is done by mouth (Alexander, 15). A small pressure cylinder of the type used for spraying has also been used to force the pollen through the tube. In Algeria, Monciero (285) has experimented with a mechanical blower of the type used for dusting sulphur and recommends it for the pollination of tall palms by attaching several sections of light metal tubing as required.

3. Factors affecting Efficiency of Pollination

a. Amount of pollen

The belief is widespread in North Africa and the Near East that different varieties of dates require different amounts of pollen (Charlet, 86; Brown and Bahgat, 62); Hughes-Buller, 206; Nixon, 329). This is usually expressed in terms of the number of strands of male flowers required for pollinating a female inflorescence; the number recommended may vary from one to 10 according to variety. Fischer (168) reported that in Mesopotamia dark-fruited varieties are said to require only a little pollen, while light-coloured ones require more. Some varieties have larger inflorescences than others, which would justify more pollen, but that does not seem to be the reason. The idea has not been subjected to scientific investigation.

On the other hand, in experimental plantings of the U.S. Department of Agriculture in Coachella Valley, California, all except three or four of more than 100 varieties of dates from Algeria, Egypt, Iraq and Baluchistan (now part of Pakistan) have been pollinated uniformly with satisfactory results.

The few exceptions could not be attributed to insufficient pollen. Probably the amount of pollen regularly used has been more than enough for any variety. There is no evidence of ill effects from too much pollen and it is considered good insurance to take no chance of applying too little. However, when pollen is scarce early in the season, growers in Coachella Valley, California, sometimes dilute it with equal parts, or somewhat more, of old nonviable pollen and have obtained satisfactory results.

Left, Barhi spikelet 35 cm long; centre, male spikelet 12 cm long; right, 50 cm measure. Ahwaz, Iran.

b. Viability of Pollen

There are known to be differences in percentage of viable pollen among different palms. Monciero (287) also found that pollen from the earliest and the latest inflorescences was inferior to that from others on the same palm. He reported further that a certain summation of heat is required to bring pollen to maturity and unless there has been this minimum amount of heat during the development of the male inflorescence, the spathe may open and pollen may be present but the percentage of germination will be low and may sometimes be responsible for a poor set of fruit. This same pollen if held for a few weeks at a favourable temperature will show a high percentage of germination. The fact that pollen may not be mature could explain some of the occasional failures to get good sets of fruit from the first male inflorescences that open in the spring.

c. Receptivity of female flowers

With most varieties of dates maximum sets of fruit will be obtained from pollinations within three or four days after the opening of the spathe, but fair sets may be obtained in some cases up to eight or 10 days (Albert, 9; Péreau-Leroy, 351). However, different varieties seem to vary in length of the period of receptivity. In Iraq it is said that the Ashrasi variety should be pollinated before the spathe opens naturally and another variety, Barban, not until about 20 days after the spathe opens. In experiments (Leding, 245A; T.W. Brown and Bahgat, 62), where the female flowers were covered with bags prior to pollination, good sets of fruit were obtained up to 10 or 12 days, and in one case there was some fruit set when pollination was delayed 30 days.

d. Weather conditions during pollination

Low Temperatures: The relation between low temperatures at pollination and poor sets of fruit has been repeatedly observed. The belief held by natives in Algeria that pollination should be done only during the middle of the day (Surcouf, 416) was tested experimentally by Péreau-Leroy (351), who reported 10 to 15 percent higher set of fruit from pollination between 10.00 am and 3.00 pm than those in early morning or late afternoon.

Laboratory tests have shown that a temperature of about 95°F is most favourable for pollen germination, which slows down as temperature drops below this peak (Reuther and Crawford, 379). In experiments at the U.S. Date and Citrus Station the set of fruit improved when pollinations were made during cool weather, if the female inflorescences were covered with paper bags. In Nejd, Saudi Arabia, early inflorescences are said to be protected from cold by wrapping them with shoots of fibre from the palm trunk (Nixon, 329).

Rain: Poor sets are often blamed on rain during pollination. Rains would be expected to wash away much of the pollen previously applied, but there is evidence that if conditions are favourable at the time the pollen is applied, pollen reaching the stigmas of the flowers begins to germinate rapidly and fertilization is not effected after a few hours. At the U.S. Date and Citrus Station, set of fruit was reduced on inflorescences exposed to drenching sprays of water

The male spathes with their enclosed inflorescences which have split or are about to do so, are collected early in the morning and the inflorescences divided into sprigs of three, four, five, six or seven spikelets, which the pollinators will use during the day. Each sprig is tied with threa‹ Basra, Iraq.

four hours after pollination, but not on those similarly treated 24 hours after pollination. Péreau-Leroy (351) exposed female inflorescences to simulated rain at intervales of two, four, six, 16 and 24 hours after pollination and found no effect on fruit set after six hours or more. On the other hand, rain six hours before pollination was said to reduce the set of fruit by one quarter or one third. Possibly the receptivity of the stigmas was reduced by contact with water. On the other hand nothing was said about the temperature, which often drops following a rain, so that the low temperature, not the rain, might have reduced the efficiency of pollination.

Wind: Severe hot wind has been observed to dry out the stigmas of the flowers, and pollen reaching them later might not be effective. However, such winds are more likely to occur during the latter part of the pollination season, at which time there is often enough pollen in the palms from early applications to effect pollination of the later inflorescences as the spathes open.

4. Handling and Storage of Pollen

Male flower clusters should be cut early in the morning, as soon as possible after the spathe breaks, to prevent wind or bees from causing loss of pollen. In fact, in North Africa and the Near East most of the main spathes are cut a day or two before they would open naturally. An experienced person can usually judge the maturity of the unopened inflorescence by pressing the lower portion of the spathe between the thumb and forefinger, which will produce a crackling noise if it is mature.

As freshly opened flowers contain a great deal of moisture, it is very important, if they are not to be used immediately to dry them out promptly. Otherwise molds may develop and destroy the pollen. When the male spathe is cut before it opens, the inflorescence is removed and usually dried out for a day or two before it is used. The strands may be cut from the male inflorescence and spread in thin layers in shallow trays in which paper can be placed to catch any pollen that falls. In California most growers collect the male clusters in the morning the spathe opens, shake what pollen they can from the inflorescences onto paper in a shallow tray for drying; then they hang up the inflorescences to dry and after several days more dry pollen can be taken from them. One enterprising grower in Coachella Valley, California, who sells pollen, has used a vacuum cleaner to extract it from the flowers.

The adverse effects of high temperatures on the longevity of date pollen have been observed in the older date-growing regions (Charlet, 86) and demonstrated experimentally in California (Gerard, 182). If exposed to direct sunlight in chambers under glass or placed near a hot stove, pollen soon deteriorates and loses its viability. However, in a dry room at moderate temperatures, pollen will keep satisfactorily throughout the pollination season (two or three months).

Pollen held over from one season to the next in storage rooms without regulation of temperature has been found nonviable in numerous tests in California (Stout, 414). On the other hand, at Tucson, Arizona, where temperatures are less extreme than in the California desert, pollen stored one year

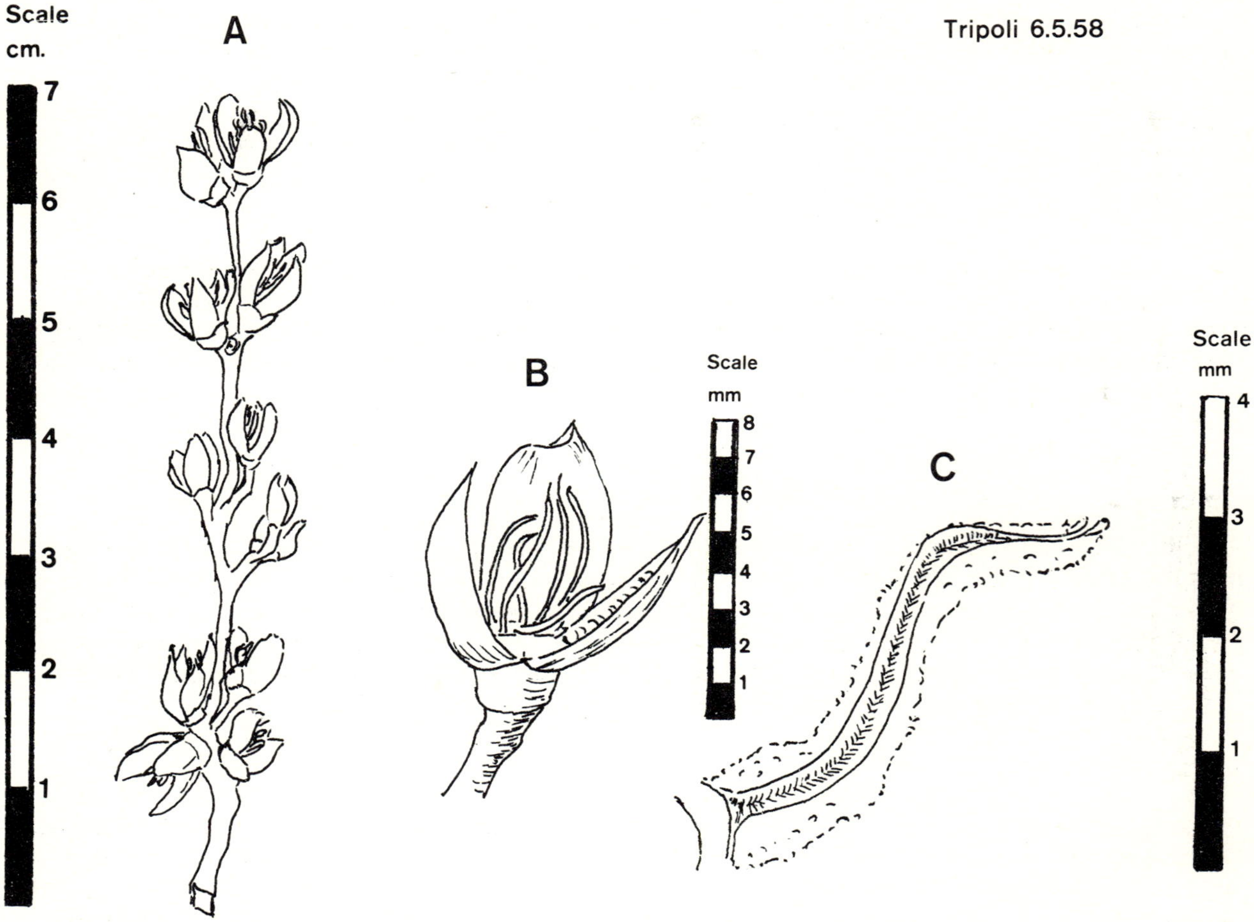

DATE PALM
MALE FLOWER

A. SPIKELET

B. FLOWER 6 stamens, 3 petals, and three-toothed calyx.
Most of the flowers have 3 petals but a few have 4.

C. STAMEN Length about 4 mm

at room temperature set fruit on 10 to 35 percent of the flowers to which it was applied (Albert, 9). These sets are much too low for satisfactory fruit production, but they are evidence that some pollen may remain viable from one year to the next if stored under better than average conditions. Cauvet (81) reported that in Algeria pollen could be held over to the following year if the dry flower clusters were kept in the spathe and wrapped in cloth.

Experiments (Albert, 9; Crawford, 99) have shown that if pollen is well dried and placed in an airtight container, it can be held in cold storage from one season to the next with very little loss of viability. This has become a regular practice in California where growers wish to be sure of having sufficient pollen early in the season. A household refrigerator at about 40°F has proved satisfactory. Somewhat lower temperatures under conditions less subject to fluctuation are probably safer. Some growers store pollen in commercial low temperature storage plants at 0°F.

5. Selection of Male Palms

Very few varieties of male palms have been developed. Three grown in the Basra region of southern Iraq are mentioned by Fairchild (153). Growers in southern California prefer offshoots of selected males, several of which are beginning to be recognized as varieties. However, most of the male palms used throughout the world are seedlings and because of the great variation among seedling palms, in selecting male palms, the grower should give careful consideration to the characters desired.

a. Time of blooming

If fresh pollen is to be used, the male must bloom as early as the females. However, unless male palms are grown under the same favourable cultural conditions as the fruiting palms they cannot be expected to flower and produce normally. Male palms are often crowded in fence rows and not given sufficient irrigation. Palms will flower earlier if planted on the south side of the garden or where there is maximum exposure to sunlight.

b. Number and size of flower clusters

With more and larger inflorescences, fewer male palms will be required. One male should furnish pollen sufficient for at least 50 female palms. Some of the best males will take care of several times that many. Under certain conditions the size and number of inflorescences may be of less importance than the effect of the pollen on the fruit, discussed later under metaxenia.

c. Flowers and pollen

Flowers that tend to adhere to the strands without shedding easily are to be preferred, and if fresh flowers are to be used in pollination, the pollen will be better conserved if the petals are not wide open when the spathe first splits. The flowers should contain an abundance of pollen. Males vary greatly

A. & B. Two (short) spikes
From different palms of the Burunsi variety

B. Four days after emergence from spathe

C. Unpollinated flower
Showing two of the three petals, the three-toothed calyx, and three unfertilized carpels

D. Unpollinated flower
Vertical section six days after emergence from spathe

E. Beiyuudhi fruit forty days after pollination. Vertical section

F. Fruit thirty-eight days after pollination
I. Brown, withered tops of unfertilized carpels, which drop off
II Transverse section. The shaded parts represent the bases of the unfertilized carpels
III. Transverse section, showing stigma, seed, remains of unfertilized carpels, corolla, and calyx

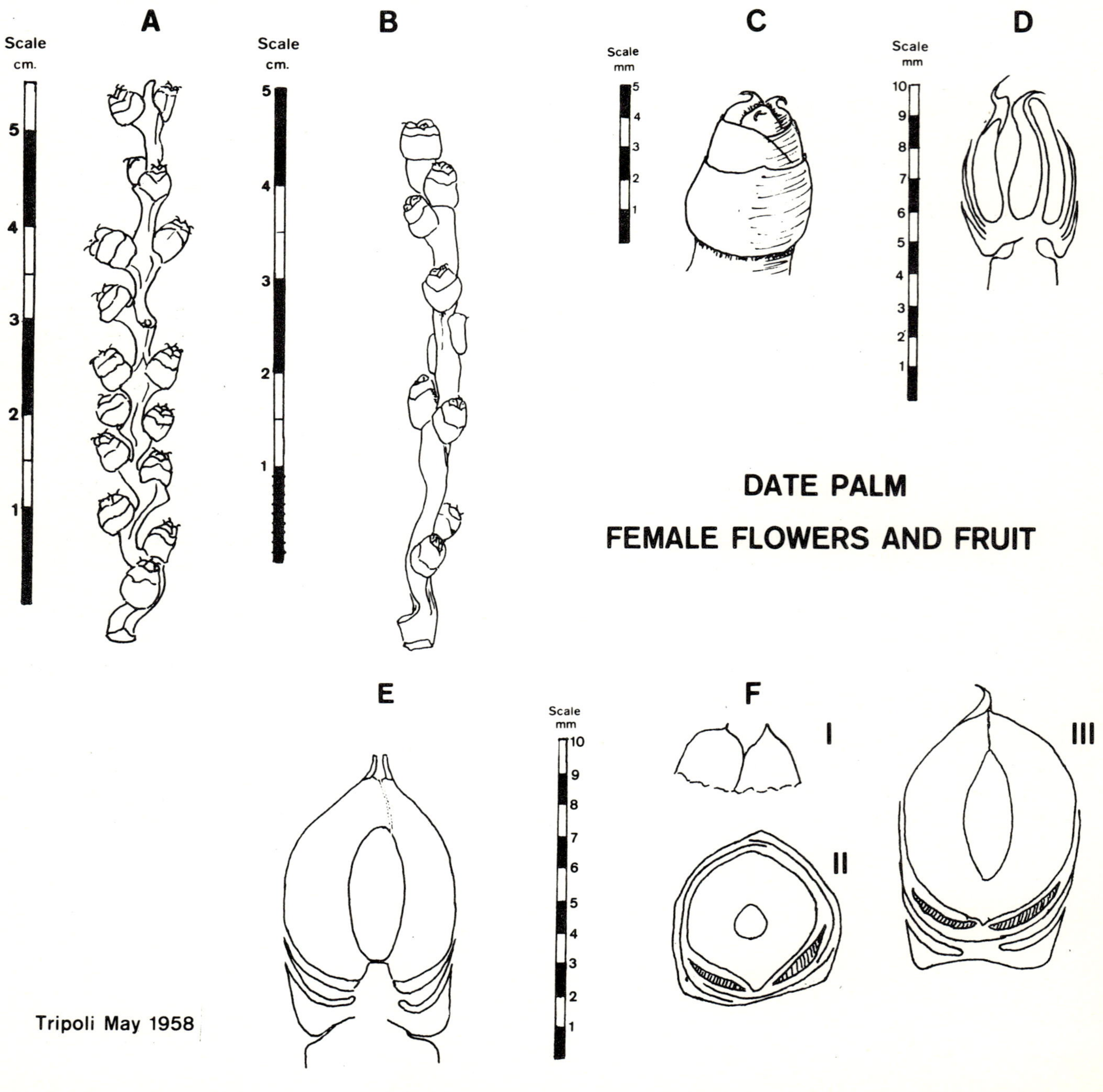

DATE PALM

FEMALE FLOWERS AND FRUIT

Tripoli May 1958

in this respect, as shown by records kept by Monciero (285) at El Arfiane, Algeria, where the production of pollen from different palms ranged from 267 to 754 g per year. At Biskra, Algeria, Wertheimer (454) reported that the average production from 12 selected male palms was 740 g of pollen per year. The highest production from one male in one year was said to be 2 133 g (4.7 lb), but the records showed that years of high production alternated with years of lower production in both number of inflorescences and weight of pollen.

d. Compatibility

The belief, widely held in many date-growing regions, that some males are better than others for pollinating certain varieties, has been borne out by observations and tests in California (Mason, 273). Only in a few varieties are the differences in set of fruit produced by different males likely to be large enough to be important, but before a male is selected it is desirable to know that its pollen will be satisfactory for the varieties on which it is to be used.

Compatibility is probably involved in the belief found in some places that better sets of fruit will be obtained with a given variety if it is pollinated by a certain type of male, usually a seedling of the variety in question. Compatibility might possibly explain the alleged "romantic attachment" of some females to certain males (Gandz, 180); also the practice in French Somaliland, as reported by Taschdjian (423), of using only males of "hot" varieties, as "cold" males are said to be less effective.

e. Metaxenia

Metaxenia, or the direct effect of pollen on the fruit outside the embryo and endosperm of the seed, is a factor that should be considered in the selection of male palms. Experiments begun in 1925 at the U.S. Date and Citrus Station, Indio, California, proved that within certain limits the pollen used affects not only the size of the fruit and seed but also the time of ripening (Nixon, 311, 312, 317, 318; Swingle, 420). Size of fruit is affected more by fruit thinning than by the kind of pollen. Size of seed is affected more by the kind of pollen than by fruit thinning and is not at present a factor of much importance within varietal limits. The most important practical application of metaxenia lies in the possibility of bringing about beneficial changes in the time of ripening of dates.

The changes in time of ripening that may result from the use of different pollens may not be more than a few days where fruit ripens early, but, because, of the retarding effect of cool weather on fruit maturity, they may be as much as several weeks where it ripens late. This was demonstrated in several date gardens in Coachella Valley, California (Nixon, 313). In one Deglet Noor garden where the fruit was not all harvested until 15 March with the pollen generally used, the application of a different pollen resulted in the completion of the harvest by the end of December (Whittlesey, 458). A few growers in that locality continue to use pollen in the same way to avoid a prolonged harvest and bring about more uniform ripening.

No differences in effect on time of ripening were found among more than 30 different males tested in Egypt (T.W. Brown and Bahgat, 62). A possible explanation is that large metaxenial differences are not likely to occur among males of any one locality where dates have been grown for a very long time because all the palms in that locality may have originated from one source and consequently may be less heterozygous than they would be otherwise. Also, if tests are made in localities where conditions are most favourable to date ripening and all varieties reach maturity with more than enough heat units, differences in metaxenial effects may be too small to be easily detected.

At Zagora, Morocco, where the climate is hot and dry up to harvest with more than ample heat units to ripen all varieties, Péreau-Leroy (351) found no difference in the effect on time of ripening of fruit among 25 different pollens applied to three different varieties. On the other hand, in a similar experiment at Ksar es Souk, where the climate is not as hot as that of Zagora, some differences in ripening were observed. Pollen which produced early ripening in experiments in California advanced maturity about two weeks when used in tests at the El-Arfiane research station in Sahara (FAO, 157). Metaxenia has also been verified by investigations in Israel (Comelli, 90) and in Pakistan (Ahmad and Ali, 3) and date growers in the former country are making use of certain pollens to ripen dates earlier in some localities.

There are without doubt, vast possibilities for the improvement of date culture in marginal areas throughout the world through the utilization of metaxenia. Wherever the sum total of heat during the period of fruit development is insufficient to ripen fruit of late varieties or where there is increasing danger of rain damage in late autumn, it may be possible to speed up maturity by the use of different pollen. Whether the crop would be benefited by earlier ripening in any particular case can be determined only by tests in which the pollen being used is compared with one known to produce early ripening. Under some conditions, such as a summer rainy season, later ripening may be desirable, and it might be possible to find male palms whose pollen would produce it.

Experiments should be started in all marginal date areas to determine whether fruit maturity can be better adapted to climatic conditions by the use of different pollens. While testing local males an effort should be made to get diverse types by means of offshoots or seeds from other countries. There is some evidence (Nixon, 317) that the metaxenial effects of pollen tend to be transmitted to seedling males of the next generation.

B. BREEDING

1. Genetics

Little is known about the genetics of the date palm, as the few breeding experiments on record have not been carried on long enough to provide much information. Some observations and a few pertinent facts that have been revealed by preliminary studies can be summarized.

Date palms vary greatly when grown from seed. All, or nearly all, varieties are believed to have originated as chance seedlings. "Nearly all" was inserted because bud variations, or what might better be termed sectorial mutations in the case of the date palm, occur and might be propagated through offshoots (Mason, 247; Shamel, 404; Nixon, 327).

Such mutations appear to be rare and the few that have been recorded have been inferior to the parent variety, but the possibility that a new variety might originate in this way cannot be ruled out.

Even though two seedling date palms are never exactly alike, seedlings tend to resemble the parent rather than some quite different variety. No one familiar with the parent varieties would have much difficulty distinguishing Zahidi seedlings, for instance, from Medjool or Deglet Noor seedlings when seedlings of each variety are seen in large blocks. Resemblances to the parent variety occur in both vegetive and fruit characters. The tendency to resemble the parent variety seems to be stronger in some varieties than in others. Although it is extremely rare for a seedling of any of the best varieties to produce fruit equal to that of the parent in quality, there is sometimes enough resemblance to cause confusion. Mason (273) called such seedlings "satellite seedlings". He described two satellite seedlings of the Hayany variety in Egypt. Natives in Iraq claim there are two types represented in both the Zahidi and Kustawy varieties. The possibility that metaxenial characters may be transmitted to seedling males through pollen has already been mentioned.

Any breeding programme is complicated by the dioecious character of the date palm. Of any large number of seedling palms about half will be male and half will be female. Furthermore, nothing is known about the male parent of any of the principal varieties. This has suggested the possibility that by following a programme of backcrossing for several generations, dates obtained from seed might begin to approach the parent variety in genetic composition (Mason, 269; Maire, 260).

2. Breeding Experiments

Early in the present century the United States Department of Agriculture distributed many date seeds for planting in California, Arizona, and southern Texas. Most of these seeds were the result of pollination by a male of the same type as the female variety and were classed as "pedigreed seeds", as distinguished from seeds produced by unknown or unselected males (Traub and Robinson, 435). Because of close planting or neglect most of the palms grown from the seeds were never really tested. Large numbers of the palms were destroyed in a campaign to eradicate parlatoria date scale (Parlatoria blanchardi (Targ.). Only a few growers here and there took care of their seedling plantings and developed them in a rational manner by selecting the best palms and culling out the others, but in this way a number of new and promising varieties have originated, of which several hundred palms have been propagated and of which fruit is being handled on a small commercial scale by the owners.

This experience emphasizes the importance of adequate planning, supervision, and study in any programme for the planting of date seeds.

The University of Arizona began a study of inheritance among Deglet Noor seedlings in 1912. The method employed was that of inbreeding by pollinating female seedlings from the best of the male seedlings of the same generation (Freeman, 170). From a study of the fruit produced by the first generation of seedlings it was concluded that the "Deglet Noor variety is homozygous for the light coloured fruit, whereas the Deglet Noor male seedling used as a parent in this cross had for a male parent some dark coloured date" and that "the dark colour of certain dates is a unit character" (Freeman and Bryan, 171). Only three seedling generations were grown before the project was discontinued. None of the seedlings produced fruit equal in quality to that of the Deglet Noor parent.

A curious result was a tendency for some of the male palms in the second generation to bear rather sparse pollen and then set fruit, showing a kind of sex reversal.

The El-Arfiane station in southern Algeria conducted a date breeding ing programme for about two decades prior to the departure of the French personnel in 1962. The objectives were to produce a line of Deglet Noor that would breed sufficiently true to type so that propagation of the variety by seed would be feasible and to produce new varieties, particularly males, that would flower early, and others that would yield large quantities of high-quality pollen (Monciero, 284, 288; Péreau-Leroy, 348). A male that flowered in January was said to have been obtained from the first backcross of the Rhars variety, but no results with fruiting varieties have been reported.

A more comprehensive breeding programme was begun by Nixon and Furr (338) at the U.S. Date and Citrus Station, Indio, California in 1948 to:

(i) produce, by backcrossing, males that are similar in genetic makeup to the parent variety;

(ii) use males from advanced backcrosses for intervarietal crosses to produce new and better fruiting varieties; and

(iii) select superior seedlings, male or female, that appear in any generation and that have possibilities as new varieties.

All available varieties that possess some outstanding character are included. Few backcrosses have been made in less than five years and more often it has been longer. However, some males of known parentage were available at the start and up through 1964 the third backcross had been made in several varieties, the fourth in one variety and some intervarietal crossing was begun. Among the backcrosses several female seedlings have shown sufficient promise to justify further testing. Results cannot be expected from any breeding programme in a short time, but it should pay dividends in the future.

C. SELECTION

1. Varietal Characters

Any breeding programme must start with varieties with known characters. Taxonomic characters of fruit and palm are important and fundamental but will not be considered here. Of more immediate concern in most parts of the date-growing world are problems of adaptation. Characters of value for adaptation can, for the most part, be determined only after careful investigation, but they are of special interest and importance in plant breeding, as will be brought out in the following review.

a. Time of ripening

The relative time of ripening of a particular variety is important in adapting date culture to marginal areas. Thus at Gabes on the coast of Tunisia, where the much prized Deglet Noor of the interior desert will not mature because of both inadequate heat and early autumn rains, there is nevertheless an important date culture based mainly on three early ripening varieties: Kenta, Aguewa, and Bouhatem (Nixon, 326). There is a similar situation at El Kantara in the foothills of the Atlas Mountains in Algeria, where date culture depends to a large extent on another early variety, Bouzeroua.

b. Rain tolerance

No variety will stand much rain or continuously high humidity, but a few, such as Dayri, Halawy, and Kustawy of Iraq, have in southern California and Arizona survived occasional rain and short periods of high humidity with much less damage than others, such as Iteema and Deglet Noor of Algeria (Nixon, 315,316, 320). The implications for some localities are obvious.

c. Disease resistance

Conditions may arise which make it imperative to obtain disease-resistant varieties if the date industry in a given locality is to survive and prosper. An outstanding example is Morocco, since the advent of the dread bayoud disease caused by the fungus *Fusarium albedinis* (Killian et Maire) Malençon. This malady, the worst known to affect date palms, has almost ruined the date industry of Morocco. The best varieties, such as Medjool, which was formerly exported in some quantity, have almost disappeared because they are most susceptible to the disease. Investigations by Péreau-Leroy (349,350) have shown that there are six varieties with a high degree of resistance to bayoud - Bou Stammi, Iklane, Taadmamt, Bou Zeggar, Takerboucht, and Bou Ijjou. By breeding and selection among such resistant varieties it may be possible in time to re-establish date culture in the striken areas.

Graphiola leaf spot, or false smut, caused by the fungus (*Graphiola phoenicis* (Moung.) Poit., attacks the leaves of all date varieties in all localities where the relative humidity is continuously very high. Recent studies

(Nixon, 330) in south Texas have shown that different varieties vary in their resistance to this disease; Kustawy from Iraq was much less affected than Zahidi from the same country. According to Dowson (*) the natives of Libya recognize varying degrees of susceptibility to graphiola leaf spot among varieties grown along the coast. By selection and breeding it should be possible to obtain varieties more suitable to marginal areas, where because of high humidity graphiola leaf spot is a serious problem. This would be a great boon to much of the Mediterranean Coast and islands such as Malta were dates are grown in spite of graphiola leaf spot.

d. Insect resistance

Pests of the date palm usually show preferences for certain varieties. The worst scale pest is *Parlatoria blanchardi* (Targ.), which attacks all the above-ground parts of the date palm including the fruit. In Algeria infestations of this scale are more serious on some varieties, such as Deglet Noor, than others, such as Rhars and Deglet Beida (Nixon, 324).

The date mites, *Oligonychus pratensis* (Banks), formerly *Paratetranychus simplex* (Banks), in California and *O. afrasiaticus* (McGregor), in North Africa and southwestern Asia, are probably the worst pests of the fruit in climates best suited to date culture. Studies in Iraq (Dutt, 137; Gazala and Bey, 181) have shown that some varieties like Khadrawy will be badly infested, whereas others like Sayer under the same conditions will be only slightly infested. According to Dowson (private communication), the mite does not attack Madgheiwa, the commonest variety at Ghadames, attacks Gharas only slightly, and attacks Deglet Noor heavily. Munier (293) reported that certain varieties grown in Assaba, Mauritania, are resistant to date mites. Resistance to date mite should be considered in breeding varieties for desert conditions.

e. Cold resistance

Near the northern limits of date culture, cold resistance is an important factor in variety adaptation. The freeze of 1937, when the temperatures in Coachella Valley, California, dropped to 13°F, afforded an opportunity to compare the cold resistance of many of the world's leading varieties (Nixon, 319). The Zahidi from Iraq was outstanding in cold resistance as compared with Halawy and Khadrawy from the same country. The Hayany from Egypt also showed considerable cold resistance, which confirmed Chevalier's (88) earlier report about this variety.

f. Drought resistance

Many localities in the desert areas have uncertain and discontinuous water supplies, and in selecting varieties for such situation it is important that they survive occasional water stress without serious damage. Certain varieties in southern Algeria, such as Azerza, Taddala, and Aouchet, have been reported drought-resistant as compared with Deglet Noor (Charlet, 86).

(*) Private communication

g. Salt resistence

The high salt content of desert soils and waters is a perennial problem. In Iraq the Sayer variety is reported to be relatively salt resistant (Dowson, 121). In Egypt, Mason (273) found evidence from palm growth that the Saidy variety is more sensitive than the Amhat to alkali in the soil. In Tunisia the Lemsi is rated as the most salt resistant variety grown on the coast (Nixon, 326).

h. Tolerance of high water table

In many localities a high water table is a problem. The Khunayzi variety on the Persian Gulf coast of Saudi Arabia will bear fruit with a water table within one or two feet of the soil surface, conditions that render many other varieties entirely unfruitful (Nixon, 329).

2. Need for Variety Studies

The discussion of some of the various adaptive characters that different date varieties have been found, or reported, to possess emphasizes the need for further study of date varieties as a prerequisite of any comprehensive breeding programme. Studies of date varieties that have been made so far are meager and sketchy (Nixon, 321), except for those of a few of the more important date varieties in the more important date-growing countries.

The need for variety studies is further emphasized by the enormous variation among date varieties which can be realized only by travelling in the different date-growing countries. The magnitude of this variation has suggested that if there was one original species of the date palm (*Phoenix dactylifera*), it has been modified by hybridization with other species in different parts of the range where it was found at the dawn of history - *P. sylvestris* in Pakistan, *P. reclinata* Jacq. and variant types in northeast Africa, and *P. canariensis* in northwest Africa (Chevalier, 88). Natural hybrids of *P. dactylifera* and *P. reclinata* (*P. senegalensis*) producing fruit of good quality have been reported in the Republic of Mali by Munier (292) who also found hybrids between *P. dactylifera* and *P. canariensis* in Spain (Munier, 297), one may expect to find great diversity in the genetic composition of the date varieties from these different regions.

In evaluating date varieties their environment must be taken into account. An outstanding variety like the Deglet Noor, for instance, if seen only on a heavy soil or in a humid climate might not impress one with the excellence of its fruit. Conceivably the original chance seedling from which this variety derived might have been overlooked had it occurred on the coast of Tunisia rather than in the desert of southern Algeria.

The reaction of a date variety to any particular environmental factor cannot usually be known until after exposure to it. Seach for a rain-tolerant variety, for example, should logically be made first in localities where rains occur during the ripening season, or for a cold-resistant variety in localities where injurious low temperatures occur.

But if none is found there, it does not necessarily follow that some varieties developed under more favourable conditions and never exposed to those adversities might not possess the desired character. Plants have been found to possess characters previously unknown until they were removed from their native habitat to a different environment that brought out the expression of the characters. For example, a pecan variety native to Texas has been found hardy in Canada (Yarnell, 468).

To evaluate environment as it affects the expression of various characters there should be, in every country where date culture is important or where there is a possibility or desire for its development, test plots in strategic locations where promising selections can be tried out under uniform environmental conditions. This is a government responsibility and requires the supervision of a trained agricultural worker who should have a knowledge of date varieties that will enable him to appreciate the potentialities of the selections tested. Any hope for variety improvement and adaptation lies in a programme of variety study, selection, testing, and breeding. The first three are necessary for the final fulfillment of the last, but all may be initiated at the same time.

3. Possibilities for Variety Improvement

The full potentialities of any programme for variety study and improvement cannot be realized without some cooperation among the date-growing countries. The project is too big.

No one country has enough diversity of material. Cooperation between the various date-growing countries is very desirable.

In a number of places where dates are grown but where economic conditions are deteriorating, at least one reason is believed to be poor varieties of dates. Even the best of the world's date varieties have some drawbacks and limitations (Nixon, 325). Harrison (195), who spent many years as a medical missionary in the Battinah of Oman, one of the world's largest date-growing regions, attributed the decline of date culture there in part to a lack of good varieties. Monciero (286) suggested the possibility of improving the date industry of Libya by the introduction and testing of new varieties. Munier (296) has emphasized the importance of variety studies as a basis for the extension of date culture. No one who has much first-hand information about conditions throughout the date-growing world is likely to disagree. A systematic programme looking toward variety improvement is necessarily a long-term project, but for that very reason the sooner it is initiated the better. The ultimate future of date culture may depend upon it.

CHAPTER VI

PROPAGATION, PLANTING, SPACING AND ASSOCIATED CROPS

A. METHODS OF PROPAGATION

There are two ways of propagating date palms: by seed and by offshoots.

Date varieties are heterozygous and do not come "true" from seed and moreover seedlings take from 10 to 15 years from sowing to come into bearing. The usual commercial method of propagation is therefore by the removal of offshoots.

However, the character of seedling palms is often rather similar to that of their mother and their use is not necessarily to be avoided in marginal areas where it may be difficult to obtain offshoots and where the fruit is required only for consumption. The variation in these seedling varieties incidentally, also offers a chance of finding types well adapted to local conditions.

In commerce of course, only well known and well liked varieties find a ready sale. There is therefore usually no alternative to the use of offshoots although the method has many disadvantages - it is difficult and laborious, hence expensive, and as each female palm only produces between 6 and 12 offshoots, which must be removed during the 4th and 6th year after planting, it is a very slow way of multiplying the crop. The sole prospect of speeding up the process of vegetative propagation seems to be through the use of meristem tissue cultures. Experimental work in this technique is referred to in a later paragraph.

B. PROPAGATION BY SEEDS

Although not usually recommended, it may be necessary to use this method for experimental or breeding purposes or where no offshoots are available. Male palms may also be produced in this way, since approximately half of the seedlings grown will turn out to be males. [However, the future tendency may be to propagate known varieties of males by offshoots (see "Pollination" p.)]

Nixon (333) recommends sowing in well aerated soil at a depth of 1 to 2 inches during warm weather in the spring. Although they could be sown in rows in a seed bed and later transplanted, it is probably better to sow in flower pots and when well established, to plant out in a nursery at about 2 m x 2 m after the third year. After the first flowering and fruiting, the females and males required may be selected from among the well grown seedling palms and planted out in the permanent site - at about 10 m x 10 m.

The seedlings should be regularly watered and fertilized throughout their growth.

Shortening of leaves prior to
removal of offshoots

Removal of soil around the
offshoots to be dug

Rooted offshoot ready to be
separated from mother palm

C. PROPAGATION BY OFFSHOOTS

1. General

The date-palm sucker, shoot, or offshoot starts from a bud in the axil of a frond, normally just below ground level. As the shoot grows, the frond base is forced away from the trunk of the palm and rots. The period of growth from the bud to a shoot of suitable size for removal from its mother may vary from 3 to 10 years. Jarvis and Jarvis (218) cleaned off all shoots from their palms by the fifth year. Nixon (333) remarks that "offshoots are ready for removal from the parent palm after 3-5 years of attachment, when they have produced roots and have themselves started to produce a second generation of offshoots". "To promote rooting, the base of the offshoot should be in contact with moist soil for at least a year before cutting. For offshoots slightly above the soil surface, this can be achieved by mounding the soil. For high offshoots, soil may be held around the base by means of boxes, but unless very valuable, these offshoots are more often allowed to become fully mature on the palm and then are placed in a nursery for rooting."

In Israel a polythene bag filled with moist wood shavings has been used successfully for rooting high offshoots (Raz, 373).

Shoots can be left permanently on the parent, but it is probable that they then reduce the mother's yield of fruit. In many places, however, notably the Sudan, where getting shoots to strike is notoriously difficult, three or four or more shoots are often allowed to remain attached, so that the plantation consists of clumps of palms. Successive shoots are then removed. If they are all allowed to remain, they continue to grow outwards from the original palm - producing large clumps consisting of hundreds of shoots, none of which produces a trunk. [Examples of these thickets are common in the Fezzan (Libya), in the Hadhramaut (Yemen P.D.R.), and elsewhere.]

In general, no green leaves should be removed from an offshoot until it is cut from the parent palm, as the growth of an offshoot will be in proportion to its leaf area. When leaves interfere with cultivation they should be tied back rather than cut off.

Irrigation several days before removal of the offshoot is recommended, so as to facilitate digging and balling.

2. Removal of Offshoots

The leaves of the offshoot to be removed will have to be tied up and trimmed. The soil is then dug away from the base with a sharp spade or digging hoe, leaving a ball of earth 2 or 3 inches thick attached to the roots but exposing the connection to the mother palm on each side. By the time the shoot is big enough to be removed, its part of attachment, the hilum, may vary in diameter from about 3 to 15 cm at its narrowest point where a clean cut should be made. The best instrument for this is the crowbar chisel as used in California. Nixon describes this as "a rectangular cutting blade made of the best highly

Offshoot being separated by
using heavy iron chisel

Separated offshoots

tempered tool steel welded to a tough iron handle". The blade is 4½" wide, 9" long and 1" thick and the handle 48" long and 1¼" thick.

One side of the blade is flat and the other beveled so as to form three sharp cutting edges. During the operation the flat side of the chisel is put toward the offshoot and the beveled side towards the parent palm. One man holds the chisel, which is driven by his assistant wielding an 8–10 lb sledge hammer. This implement is fairly widely used now by the wealthier date growers in North Africa and the Near East, although often smaller in size than the Californian version. In Iran it is often used without the hammer - the chisel being raised with two hands and brought down with heavy thuds on the hilum until the shoot has been separated. Poorer growers are often not able to obtain a crowbar chisel but manage to achieve a clean cut with a sharp axe. In the Fezzan (Libya) a small "back cutting" saw is used, but this is not recommended as a general practice.

Great care in these and succeeding operations is required, as mortality of offshoots is often high and represents considerable loss and inconvenience to the planter. To obtain a good clean cut through the hilum is the first requisite. In Israel it has also been found useful to dip the lower part of the shoot before planting in a weak solution of copper sulphate, to protect the cut surface against Diplodia attack. Cut surfaces of fronds not reached by dipping are brushed with the same fungicide. In 1965, Bulit and Gerard (66) reported experiments made at El-Arfiane in Algeria with various fungicides brushed on to the bases of offshoots after removal from their mothers, from which it appeared that the most promising treatment involved the use of an aequeous paste of 60% Triphenyl-tin-acetate with 20% Manebe. The general opinion of experts is that the best results are obtained with shoots weighing at least 18 to 20 kg after trimming. Whenever possible, offshoots of this size should be selected and the small shoots left on the parent plant for a few years longer.

There are numerous records of offshoots being planted of anything from 1 to 50 kg in weight, in different countries [7,453] but it is not clear what is the precise relationship between size, survival and growth.

3. Treatment After Removal and Transport of Offshoots

Nixon (333) recommends cutting off the lower leaves and old leaf stubs close to the fibre, leaving the basal 2 to 4 ft of the offshoots bare of leaves. "Ten or twelve leaves around the bud are retained and tied close together a few inches above the bud with heavy twine or wire. The terminal parts of these leaves extending beyond the tie are also cut off."

Offshoots frequently have to be transported long distances and in view of the difficulties often experienced in rooting, great care should be taken of them from the moment of their removal. Drying out the shoots in the shade has been recommended in the past, but the practice has been discredited. All efforts should be directed at preventing the roots from drying out. Although roots cut during the operation of removing the shoots apparently mostly die, it is essential to avoid damage to the small root primordia. For this reason, if the shoots cannot be planted immediately, they should be shaded and the roots covered with

Offshoot ready for planting

moist soil, and if there is to be more than 1 or 2 days' delay before despatch, they should be planted in a temporary nursery and kept irrigated. In the U.S.A. balling with wet sphagnum moss and burlap is often practiced with offshoots that have to be shipped some distance. In Africa and the Near East, date fibre ("leef") and old sacking is used. If the shoots are transported by river, it is a simple matter to keep the packing moist, but when carried by train, lorry or camel it is necessary to go to some trouble to prevent them drying out.

During transport rough handling should be avoided, as it is possible for the bud and tender heart leaves to be damaged by being dropped or subject to undue strain. [333]

4. Time of Planting

In most of the date lands, spring and autumn are preferred for the planting out of date shoots. Spring avoids the cold of winter and takes advantage of the warm weather that causes rapid growth, while autumn gives the young shoot a longer time to establish itself before the heat of summer; but each season has its corresponding disadvantage; spring, the early approach of the great heat, and autumn, the early approach of the cold. Probably date shoots can be safely planted at any time, except during the hottest and coldest periods of the year, provided irrigation is plentiful and frequent enough. June planting has proved successful in Israel.

On the Nile, however, the season preferred for the planting of shoots is August and September, just after the hottest part of the year, probably because it is only then that the river comes down in flood, making plentiful and frequent irrigation easy.

Al Bakir and Al-Azzawi (7) record the results of a date of planting experiment with Zahidi shoots at Baghdad (Iraq). One hundred shoots were planted every fortnight throughout the year. The best results were obtained from the eleven plantings made from 15 April to 15 September, 79% survivals. In the remaining thirteen plantings, 1 October to 1 April, there were only 37% survivals. The highest survival rate was in the 1 June planting, 95% (lowest minimum temperature 17°C) and the lowest survival rate, 8% (lowest minimum temperature 7°C) was that of the 15 January planting.

It is to be presumed that, as the Zahidi is the commonest variety in Baghdad, there was no difficulty or delay in procuring the shoots, so that the time between removing the shoots from their mothers and planting them was short. It is the general view that delay before planting should be as short as possible.

5. Planting the Offshoots

In some countries where, as in the Sudan, there is a great danger of losses after planting, it is found convenient to establish offshoots in a nursery near to the planting site before final planting. This obviates replanting of dead shoots and subsequent uneven development of the plantation.

Offshoot being planted

Freshly planted offshoot
soil being firmed

It is important that the soil in the nursery bed is well drained and kept well watered. A light shading of palm fronds is sometimes provided.

In other countries, such as Iraq and Iran, where less trouble is experienced with establishing offshoots, they are planted direct in their cropping sites. Nixon (333) recommends planting holes of about 1 m diameter and 1 m deep. "They should be filled with a mixture of topsoil and barnyard manure and subsequently irrigated several times to promote decomposition of the organic material" (i.e. before planting). When planting in good silt soil, for example in the Nile Valley, this is not considered necessary and the hole is dug just deep enough to accommodate the offshoot, but when planting in very heavy or stony soil it is often thought better to prepare a planting hole of about one cubic metre in advance. Where there is a hard pan of rock or gypsum or of sandy gypsum crystals near the surface, this should be broken through if it is thin enough. If not, the palms should be planted elsewhere. Enough time and irrigation must always be given to allow the soil "refill" to settle in the pit before planting. The level of the earth after watering and drying out should be that of the surrounding soil, otherwise the offshoot may be lowered too much after irrigation.

It is important that the offshoot should be planted and maintained at the correct depth, i.e. at the level of its greatest diameter, and never so deep that the irrigation water will reach the loose fibre round the growing point. Soil should be carefully tamped down round the offshoot so that the base is kept in close contact with the soil at all times. A basin 15 cm to 30 cm deep and 1 to $1\frac{1}{2}$ m in diameter should be formed round the shoot. After planting it should be given water immediately and then irrigation is usually given every 3 to 4 days. Nixon warns however "it is essential that the soil near the newly planted offshoot be kept moist at all times by light, frequent irrigations. Inspection should be made often during the first few weeks to see that the surface soil does not dry and shrink away from the offshoot".

"The frequency of irrigation will depend somewhat on the soil type. During the first summer, daily irrigation may be desirable on very sandy soils. Every second or third day will not be too often on moist soils, but on very heavy soils, once a week may be sufficient".

"For protection against sun and wind during the first summer and against cold the following winter, the newly planted offshoot should be wrapped with burlap or with a layer of leafy material such as corn stalks or date leaves. The top should be left open so that the new growth may push out".

6. Mortalities

If all the above precautions are taken, a good percentage of success may be expected, but it must be remembered that in some countries difficulties in rooting are common. In the Northern Sudan, for example, this may be due to the low atmospheric humidity (between 9% and 20% Mean R.H. at 1400 hours); and high temperatures (mean daily maximum 30°C to 43°C). The lowest humidities usually

One metre-cub holes being dug. Ahwaz, Iran.

coincide with the highest temperatures.

Dowson reports an unfortunate experience in Northern Somalia in 1950 to 1952 where deaths of from 50% to 90% were suffered in a collection of varieties imported from Hadhramaut (People's Democratic Republic of Yemen), Masqat, Bahrain and Sudan. Indeed the only good results (76% survival) were obtained with offshoots imported from Tunisia and first planted in a nursery in the hills at Hargeisa. Smead was more successful in importing 2,000 offshoots from Pakistan, to the north of Kenya in 1971 (*). He reported survival rates 6 months after planting of 4 female varieties varying from 57% to 100%. All experts agreed that, when importing offshoots from a distance, the planter should supervise the selection, removal and packing of the offshoots himself and stay with them until they are properly planted. In this way he can ensure that the work is done properly, that the roots are kept moist and that selection is made of the right type of offshoot, from the correct variety. Importing large numbers of offshoots from another country is in any case no small undertaking; incidentally, a considerable diplomatic effort is often involved!

Dowson gives some notes on experiences personally known to him or reported from various countries:

Lyallpur (Milne, 277): 50 date shoots were planted in 1910, 43 were alive in 1916, survival rate 86%. 50 shoots were planted in 1911, 42 were alive in 1916, survival rate 84%.

Coachella Valley, California (Cook, 92): 15 well rooted Deglet Noor shoots were planted in 1920, all survived in 1924. Of 130 planted in 1921, 98% survived. 171 planted in 1922 showed same survival and same survival was expected of 264 shoots planted in 1923.

Basra, Iraq: 2,878 Hallaawi shoots planted on a large estate between 1927 and 1930 showed a survival of 61% in January 1931.

Somalia (Dowson, 125): Between November 1950 and November 1952 some 4,000 shoots from different countries were planted in test stations. In February 1954 they showed a survival of 27%.

Libya (Dowson, 126): Between mid-March 1958 and July 1959, 646 shoots of 66 varieties from Western Libya were planted in the Guea Date Varietal Test Station. 79% were alive at the end of 1959.

Ziban, Algeria (Wertheimer, 453): Here 20% to 30% usually die in the first two years from planting and the general survival rate is estimated at about 60%.

Israel: As from 1948 the earlier plantings showed rather high mortality but in more recent ones it has been extremely low, viz. less than 5%.

(*) See FAO Report AG:DP/KEN/73/002. 1976

Dowson remarks that Wertheimer's figures for Ziban are probably valid for most of the Old World countries except Israel. Both in Israel and California deaths are fewer. It may be noted that in these two places shoots were very hard to get in the early days of the industry and hence great care was taken in separating the shoots from the mother plants.

Wertheimer's conclusions were:

a. The mother should be healthy;
b. The shoot should be young, in a state of rapid growth, big enough, and roughly cylindrical;
c. Shoots must have put out roots;
d. Only skilled operators should cut through the hilum;
e. The surface of the cut should be small, smooth, and without cracks or fissures;
f. Plant from February to May. Early planting is better than late;
g. Plant as soon as possible after cutting.

From his monthly observations, lasting for a year, of the root growth of 140 newly planted shoots, Wertheimer concluded that the initial state of the cut surface and its subsequent behaviour were the most important factors influencing the survival of the shoot. He adds that the putting out of roots, the persistence of the green colour of the foliage, and even the growth of the fronds, are not certain signs of the shoots having taken.

D. PROPAGATION BY MERISTEM TISSUE CULTURES

In recent years vegetative propagation by "in vitro" techniques has proved of great value for a variety of crops, especially for the production of plant material which is virus-free as well as true breeding. It has been successful for the latter purpose for a variety of root and tuber crops, including potatoes, sweet potatoes, cassava and yams, and for the production of virus-free clones of strawberries. It has also found important commercial application for the propagation of orchids, carnations, chrysanthemums and asparagus.

There has been considerable interest in extending the use of these techniques to tree crops, but this has proved to be difficult. The most successful example concerns the oil palm, Elaeis guineensis and its hybrids, work on which started in the late 1960's in the U.K. laboratories of Unilever. Callus cultures were produced from crown and root meristem tissues and Jones and his co-workers (*) finally managed to produce plantlets which were sent by air to Malaysia. Field trials will obviously take time, but there are good prospects of remarkable increases in oil yield per hectare from the use of such clonal material in the future (*).

(*) Vegetative propagation of oil palm via tissue culture. R.H.V. Corley, J.W. Barrett and L.H. Jones. Malaysian International Oil Palm Conference, 1976.

Tunisia - This plantation of date palms has been neatly planted, with the trees widely spaced and properly pruned.

Work on date palms has not reached a comparable stage. Reuvani, Adato and Lilien Kipnis of the Volcani Institute, Rehovot, reported in 1972 (383) that, of the various tissues and organs tried, only embryos could be cultured "in vitro" successfully. In all other cases, there was a "discharge of a yellowish-brown compound into the medium; this substance inhibited the development of the tissues". (383).

If these difficulties are eventually overcome, the results for date-palm culture could be extraordinary. Large-scale and rapid multiplication of selected varieties might become possible and incidentally, may change the perspective for breeding new varieties.

It may be noted that Eeuwens and Jennet Blake of Wye College, U.K., reported in 1977 (Actae Horticulturae, 1978) considerable progress in their work on tissue culture on coconut and that "in the date palm, roots (had) been initiated on young leaf petiole tissue and successfully sub-cultured." (**)

Allowing for the slow development of date palms and for the usual long delays before a major technical agricultural breakthrough finds practical application, there seems good reason to hope that, in a few decades, clonal multiplication of date palms through tissue culture may become a standard practice. Specialists in date growing countries will, no doubt, be taking an active interest in the subject.

E. SPACING

Ibn Sidah, who died in 1055/6 A.D., recommended planting palms sufficiently far apart for the midribs of the fronds of one palm not to touch those of another. Nixon, who died in 1975 seems to have been of much the same opinion, since he reported that spacing of 30 ft x 30 ft has generally given the best results in California, except for relatively small varieties like Khadrawy which "can be planted 2 to 4 ft closer without undue crowding". Dowson points out that the chief climatic disadvantage in the Coachella Valley is humidity in the date-ripening season and that therefore rather wide spacing is desirable.

In normal irrigated conditions the highest yield will presumably be obtained when none of the crown of the palm is in shadow. At this spacing the roots should also be able to obtain sufficient water.

However, in exceptional conditions such as the humid coast of Libya, round Zliten, wider spacing is given, presumably to allow sun and wind to counteract the influence of the humidity. There are palms often planted at 6.7 to 8 m in the rows, but the rows may be 20, 30 or even 40 m apart. Wide spacing is also the rule in Elche (Spain).

(**) C.J. Eeuwens and Jennet Blake. Culture of Coconut and Date Palm Tissue with a view to Vegetative Propagation. Acta Horticulturae, <u>78</u>, 1977.

Nixon notes that wide spacing is important whenever there is considerable danger of rain damage to dates during the ripening season. The better ventilation and insolation of the palms will reduce loss from fruit rot. On the other hand, where the air is dry, and especially where the prevailing wind is not only dry but extremely hot and strong, the tendency is for date palms to be planted more closely, as at Basra in Iraq. The average density of planting in Basra is given in Table 11.

TABLE 11

AVERAGE DENSITY OF PLANTING

BASRA, IRAQ

Kind of Tree	No. per acre	No. per ha.
Female date palms	122	301
Male date palms	3	7
Offshoots	15	37
Total palms	140	345
Fruit trees and vines	39	96
Total palms and fruit trees	179	441

Dowson comments that several visiting horticulturists have criticized the closeness of planting in Basra, and indeed some of the gardens are much too crowded, but whether on the whole there is too much crowding is not yet proved, because in the years when the scorching northwest wind from the desert blows late and is not supplanted by the southerly drift from the Persian Gulf just as the Hallaawi crop is completing its ripening, many of the dates of this variety become Abu Khusheim, that is, hard and dry, and less acceptable to the merchants. Also the date palms of Basra are on the whole small, considerably smaller than those of Baghdad, due probably to the very high water table of their habitat. Seen from the air, the date palms in Basra do not appear to be unduly crowded, except in the Abul Khasib district which is, or at least was, producing the best dates.

In Algeria in 1922 the number of palms in Arab gardens was generally rather less than 100 (per ha) and in the gardens of the colonists 75 or less

(Dowson 122). Rème (378) gives the average density of planting in the Oued Rhir as 129 palms to the hectare (52 to the acre).

The Sudan Government recommended 76 to 80 palms per acre instead of the usual 100 (Bacon 27). Dowson (124) counted the number of clumps in a garden at Duwein, near Merowe, and found that there were 133 to the acre. The number of stems to the acre was 307, averaging 2⅓ stems to a clump.

TABLE 12

COMPARATIVE TABLE OF SPACING DISTANCES

(Palms planted at the corners of squares)

Distance Apart		Square Units to each Palm		No. of Palms in Each	
Feet	Metres	Feet	Metres	Acre	Hectare
33	10.06	1,089	101	40	100
30	9.14	900	84	48	119
29	8.83	838	78	52	129
28	8.53	784	73	56	137
27	8.23	729	68	60	148
26	7.92	676	63	64	159
25	7.62	625	58	70	172
24	7.32	576	54	76	185
23	7.01	529	49	82	204
22	6.71	484	45	90	222
21	6.40	441	41	99	244
20	6.10	400	37	109	270
19	5.79	361	34	121	294
18	5.49	324	30	134	333

F. ASSOCIATED CROPS

These may be considered in two groups: 1. Interplanted Tree Crops and 2. Interplanted Vegetable and Fodder Crops. The use of cover crops and green manures is discussed in Chapter VII.

1. <u>Interplanted Tree Crops</u>

Interplanting with tree fruits, especially citrus, is common in many

Jahrum, Iran.   Orange Trees under Palms

countries of North Africa and the Near East and is occasionally practiced in the U.S.A.

Under Californian conditions, which are comparatively humid during the date ripening season, the practice is not recommended. Nixon comments "if the two crops (date and citrus) together do not have at least as much space as each planted separately would require, yields of both are almost certain to be reduced. Serious checking, blacknose and rot of Deglet Noor fruit often result from increased humidity when the palms are not high enough, so that the bunches hang well above the tops of the interplanted trees".

However, conditions in the Old World often favour interplanting. In the first place, conditions are sometimes very severe for citrus, strong sun and very hot dry winds in summer and/or very cold winds and low night temperatures in winter. The palms often seem to improve the yield of citrus, which is the more valuable crop. The practice is popular for example, for sweet oranges in Baquba, Iraq, and for limes and sweet oranges in Southern Iran. It is not clear however, how much shade, if any, is beneficial to citrus and what part other factors play.

In the second place, interplanting is frequently favoured by the poor farmer who has inherited a small date garden and plants some fruit trees under the palms to provide an alternative source of income and some fruit for his family. Species which may be found mixed with palms in different districts include the apple, apricot, banana, all species of citrus, fig, guava, Indian almond, jujube, mango, mulberry, pomegranate, prickly pear and Sebesten (Cordia myxa).

Over-crowding, however, is very common. Frequently citrus and other fruit trees are planted in a date garden which already has too many palms to the hectare. The yield of both would be much improved if citrus and dates were planted in separate, possibly double or treble, rows at adequate spacing. The optimum spacing would have to be based on local observation. Naturally, in climates ideally suited to citrus, or other species, they are best grown unmixed.

2. Interplanted Vegetable and Fodder Crops

The arguments for inter-cropping under conditions in the Old World are especially strong in the case of vegetables and fodder crops. These crops are needed throughout the year in oases and in irrigated areas within easy transporting distance from towns - and yet often do not grow well during the hottest seasons. The date palm with its fronds turned edgeways to the sun, gives only half shade and this half shade and the partial shelter from the wind afforded by the palms allow cultivation of a variety of crops which could not otherwise stand the harsh desert climate. Economically, also, the practice is valuable to the cultivator, since a succession of vegetables and fodder can supply an income throughout most of the year.

When the palms are young, the best crop in Old World countries is alfalfa, which does not require shade and yields well under the hottest conditions, provided it is well irrigated. When the palms are fully grown, the alfalfa is best replaced with other crops that stand the semi-shade.

These practices are usually beneficial to the growth of the date palms for human as well as technical reasons. Not only do the roots and residues of the intercrop improve the physical properties of the soil and the greater spreads of irrigation improve soil and atmospheric conditions, but the presence of a crop giving regular cash returns encourages the cultivator to irrigate and manure his date garden regularly!

CHAPTER VII

SOIL REQUIREMENTS AND THE USE OF ORGANIC MANURES, MINERAL FERTILIZERS AND GREEN MANURES

A. SOIL REQUIREMENTS

In principle, the date palm has similar soil requirements to other cultivated crops, i.e.:

(a) sufficient depth to enable the roots to anchor the palm firmly and to prevent its being blown over;

(b) a texture such that the palm roots can ramify in it easily and that air can penetrate into it;

(c) a content of nutrient elements known to be necessary for plant life and not absorbed from the air, viz. boron, calcium, chlorine, cobalt, copper, iron, magnesium, manganese, molybdenum, nitrogen, phosphorus, potassium, sodium, sulphur and zinc.

Furthermore, the soil should not contain excessive quantities of the carbonates, chlorides and sulphates of calcium, magnesium and sodium, nor quantities higher than a few parts per million of certain other elements such as boron, molybdenum and zinc. The soil should contain sufficient moisture to enable the palm roots to absorb its nutrients in a suitably weak solution.

As the date palm grows in a climate of little rain, but great heat and much evaporation, irrigation or flood water evaporates quickly, and its salts are left at the surface of the soil. The water, moreover, which is used in irrigation in the arid regions of date palm cultivation often comes from rivers that flow through salt rock or salt pans and is consequently already saline.

Before planting in salty soil, if there is enough irrigation water for the purpose, and if the lie of the land permits, the salt should be leached away. If, however, the irrigation water available is insufficient or if there is nowhere for the drainage water to go, then the top, salty layer of soil should be scraped away.

Roots of date palms in torrent bed at Kashit, Iran, exposed owing to eating away of bank by floods. It can be seen that the development of the root system of the date palm is very extensive at a depth of between two and three metres below the surface of the soil.

The soil requirements of the date palm have not been the subject of much exact study, but many authorities on dates have given their opinion from a comparison of the results of date culture in different soils. A representative selection of opinions is given below.

Pliny (c. A.D. 23-79) says (in Holland's translation of 1601): "Date trees love a light and sandy ground".

Cosson and Hamin (1885), writing of the Ziban in Algeria, state: "The nature of the soil appears to be of less importance (than climate); "since cases are found both on the calcareous and on the gypseous clays that cover most of the region and both in saline and in sweet soils".

Bunce (1860), writing of the oasis of Khabis (now called Shahdad) in Iran, stated that the soil was a firm loam mixed with some sand and contained some salt.

Fairchild (1903), writing of Basra in Iraq, states: "In Bassorah there is an adobe soil, resembling the silt of the Nile Valley, so sticky that it has only to be dried in the sun to make the best of adobe bricks. This pure adobe is not considered as suitable a soil as that which has a slight admixture of sand in it to make it lighter".

Rivière and Lecq (1914), writing of the Sahara, state: "The land which suits the date palm best has a light soil, not exhausted by previous crops, and is well drained, either naturally or artificially, so that excess irrigation water is removed and there is no stagnant water".

Fenzi (1916), writing of Libya, says: "Apart from any consideration of the effect of a high or low water table, the best land for the (date) palm is loose and sandy, moderately well supplied with humus, potash and other nutrients. In very calcareous, compacted, and rocky land the palm is slower to develop, and will never bear abundantly, perhaps because its roots, being compelled to grow very near the surface, are too exposed and suffer from dessication".

Milne (1918) states: "The date palm appears to grow and produce fruit almost equally well on sandy, loamy or clayey soils. The soils of the Punjab plains are alluvial varying from light sands to stiff clays, and often contain a large percentage of lime"; and, later: "The physical character of the soil appears to have little

The soil auger in use. It is important to know what is under the surface of the soil before planting date palms. Qatif, Saudi Arabia.

influence on the plant, except that the flowers and fruit may be a little earlier on sandy soils. The soils of the Saharan cases are mostly of a sandy nature; those of the Mesopotamian tract and in Egypt are mostly of very dense clays, while some of the date groves on the coast of Egypt are growing on almost pure sand. On the whole, sandy loams are best suited for date cultivation".

T.W. Brown (1924), writing of the Arab Republic of Egypt, states: "Dates can be grown profitably on any variety of soil. The chief centres of production in Egypt are situated on sandy land. This is due to the fact that the trees grow as well as, or better than, anything else, on this kind of land, but it does not necessarily imply that the tree grows better on sandy land than on rich alluvial land. In fact, they appear to grow most strongly on deep porous soil such as that formed by the coarser deposits of Nile silt, and which is known as gezira or tami land in Egypt. With regard to varieties, there is no apparent difference in their soil requirements. The Samâni, which is grown almost entirely on pure sand in the Rosetta and Idku districts, thrives and bears equally well when planted on black land in other districts. The same remark applies to other varieties".

Jarvis and Jarvis (1941), writing of their own land in the Coachella Valley, California, state: "Our soil is well adapted to date culture on the average, varying from blow sand to silt. The best production always comes from the lighter soil after it has been built up. Poorer quality fruit and a low yield is always the case on heavy type soil".

Nixon (1959), writing of the U.S.A., states: "Dates are grown on a wide variety of soils. The maximum water-holding capacity consistent with good drainage is desirable. Coarse sand requires excessive fertilization and irrigation and permits rapid leaching of mineral nutrients unless underlain by more retentive soil of finer texture somewhere in the first six feet. On the other hand, good growth and production cannot be expected unless the soil takes water readily to a depth of six or eight feet. Some of the finest date gardens in southern California are on deep sandy loams".

Ochse et al (1961) state that: "The date thrives in almost any type of soil, especially in well-drained sandy loams".

The foregoing shows general agreement that the date palm will tolerate wide variation in the type of soil in which it will grow satisfactorily. However, it is doubtful whether the date palm will do well in a very sandy soil, unless it is supplied with fertilizers and one of the three following conditions is fulfilled:

(i) the palm is growing close to an almost unlimited water supply, such as an unlined and usually flowing irrigation channel or a spring;

(ii) it is growing in sand overlying an underground water course; or

(iii) it is growing in sand under which at a few metres there is a layer of soil relatively impermeable to water.

The first condition is fulfilled in date gardens in sandy country, where the palms near the well and main channels are higher than those on the perimeter of the garden; and the third condition in the numerous gardens on the sandy banks of torrents, where there is often a band of alluvium under the sand on either side of the torrent bed (see sketch by Munier, 295).

Fulfilment of condition (ii) is seen at the Wadi Zellaf in the Fezzan of Libya where, although on the surface there is nothing but sand stretching as far as the eye can see, a few metres below it, the depth varying with the configuration of the dunes from perhaps 2 m to 10 m, there is slowly flowing water. Along the bed of this water course there is a straggling line of perhaps more than 30 000 date palms, in excellent condition, and yielding well, although mainly seedlings.

It is difficult to believe that the extremely stiff clay of the Basra (Iraq) date gardens is an ideal soil. In the spring, when the river is high, the dry summer and the sodden clay is baked into a condition as hard as a brick. It must be difficult for the roots to penetrate such hard soil. Digging to break this extremely hard ground does, indeed, do great damage to the roots of the palms, which in Basra are shallow, because the water table is so high. An admixture of sand, as Fairchild mentions, makes the Basra soil easier to work and is almost certainly better for the palms. It may be noted that at Qatif, in northeast Arabia, sand from the desert just outside the gardens is regularly brought in to improve the texture of the silt and clay.

To sum up, it might be said that a light soil is probably more suited to the date palm than a heavy one provided that irrigation is frequent, or that there is a sufficient underground source of water or a more or less impermeable subsoil to hold the water not too deep for the roots to reach it.

Judging by the places where the date palm grows well, one may say that it is more tolerant than other fruit trees of soils with a high content of lime, gypsum, sulphur, and of the usual salts of saline soils.

B. THE USE OF MANURES AND FERTILIZERS

1. General

Various investigations have been made into the removal of nitrogen, phosphorus and potassium from the soil by the date palm and its crop.

For example, Monciero (287) states that the "Société algérienne et industrielle du sud algérien" informed him that an analysis made by the "Station agronomique de l'est" on 10 July 1903, showed that from one ha (120 palms), the following amounts of plant nutrients were removed annually from the soil in the midribs, spines, leaflets, fibre, fruit bunches, and 45 kg of dates (but not including the amounts in the new growth of trunk and roots): 72 kg N; 5 kg P; 27 kg K.

Haas and Bliss (191) find that the average yield of a Deglet Noor palm in southern California, 9 000 dates, contains 239 g of nitrogen, 41 g of phosphorus, and 585 g of potassium. At 48 palms an acre, the losses in these elements from one ha (120 palms) would thus be 29 kg N; 5 kg P; 70 kg K.

Embleton and Cook (150) calculated that the average annual prunings from one ha of Deglet Noor palms, say 120 palms, would contain 25 kg N; 2 kg P; and 74 kg K.

Furr and Barber (177) estimated the average annual loss of nitrogen from the land in the Coachella Valley bearing Deglet Noor palms, presumably spaced at 120 to the ha, at 78 kg, while samples of soil taken from both inside the date gardens and the virgin desert outside showed that the nitrogen reserves in the top 2,4 m of soil varied from 2 268 to 17 010 kg per ha.

It is assumed that these enquiries were made at least partly in the hope that they would throw some light on the manurial requirements of the crop. However, the soils of arid zones frequently contain high reserves of P and K, as well as of other minerals, and hence the results of applications of fertilizers to date palms have frequently proved inconclusive or non-existent.

Applications must be based on field trials and observation and not on the weight of elements removed by the crop. In practice the subject is often complicated by the needs of interplanted crops and cover crops and by the need for the physical improvement of many soils. The relative advantages of organic manures, "green" manures and inorganic fertilizers must be considered.

2. Organic Manures

Animal manures are widely used in the better date gardens of the Middle East and North Africa, although in most of the poorer ones no manure is used unless it is for the vegetable and fodder crops grown between the palms.

Dowson notes that the traditional practice in Basra date groves, where there is almost no surface irrigation, was to bury animal manure in deep circular trenches around the palms where it could be reached by the subsoil water. In desert oases, the dung of sheep, goats, camels and donkeys is widely used and in Egypt, along the Nile, large quantities of pigeon dung, barnyard manure and night soil are used, or buffaloes are tied to the palms for a few days. Surcouf (416) notes that camel dung is the chief manure of date palms in Algeria and that it is put into a deep trench dug all round the palm. [Incidentally, he suggested that it would be an improvement if each year only a quarter of a circular trench were dug, so as to save the roots]. Johnston (220) dives a useful table showing the usual percentage composition of various organic manures (Table 13), in California.

The quantities of animal manure used naturally vary a great deal. Five to 15 tons per acre is the range of applications mostly reported from California and Arizona - usually made in the late autumn or winter. If a winter cover crop is grown, manure is usually applied in the spring after the cover crop has been turned under. There can be little doubt that such practices are beneficial to soil fertility but, under economic conditions in Old World date gardens, ploughing-in crops of any sort is not likely to be a sound proposition.

Table 13

PERCENTAGE COMPOSITION OF MANURES AND STRAWS

(Johnston 1944)

Manure or straw	N	P	K	Organic matter	Water
Dairy	0.80	0.45	1.37	30.5	32.70
Steer (cotton seed fed)	2.00	0.55	1.92	58.0	14.50
Steer (alfalfa fed)	1.35	0.65	2.70	48.0	16.80
Poultry (droppings)	4.15	3.13	1.58	74.0	8.27
Poultry (manure)	2.00	1.85	1.16	52.0	12.50
Rabbit	2.25	1.35	0.83	60.0	5.50
Hog	2.20	2.10	1.00	62.0	10.00
Sheep	1.40	0.95	2.10	52.5	9.00
Lima bean straw	1.20	0.25	1.28	82.0	8.50
Blackeye bean straw	1.00	0.25	1.90	82.0	8.60
Grain straw	0.60	0.30	1.38	86.0	5.79

3. Mineral Fertilizers

As there has been little experience of the use of mineral fertilizers in date gardens of the Old World, conclusions must be drawn mainly from reports from the U.S.A.

Nixon (333) pointed out in 1966 that there were few real experimental data bearing on the kind and quantity of fertilizer to use or on the time of its application; however, some trials have been carried out and there are some indications of the elements most often required and the quantities needed. As Nixon (333) says: "on most soils in arid regions, tree crops have failed to respond to applications of potash and phosphorus". In fact, the main interests in fertilizer applications for dates in California is concentrated on Nitrogen.

Furr et al (179) report the results of seven years' manurial trials with Deglet Noor palms. During the first three years the application of nitrogen had no apparent effect on growth or yield, but during the last four years the manured palms had rates of trunk growth, inflorescence production, and yield one-fifth greater than those of the unmanured palms. For the first three years, each manured palm received 6 lb of nitrogen a year and this was increased to 8 lb the last four years. The soil map showed the soil as Coachella fine sand, but in much of the garden there were layers of silt and clay of different thicknesses in the top ten to 12 ft.

The value of nitrogenous fertilizers is borne out by other field trials. The following extract is taken from Bliss and Mathez (42) on the Arkell date fertilizer experiment in California:

> "Despite the original high level of soil fertility in the experimental plots, and the addition of nitrate by the irrigation water, the response of the palms to additional fertilizer (on the basis of comparison with the average performance of palms in the four control plots, 1939-1942) amounted to about 46 lb of fruit (gross return) and six inches of trunk elongation per palm a year. This response is thought to be mostly due to the addition of nitrogen. The data are not considered to be sufficiently critical to indicate whether or not the additions of potassium and phosphorus had any significant effect. There is also little or no experimental evidence that a response was obtained from the application of organic matter to this soil."

The irrigation water supplied 5.15 lb of nitrate, i.e. 1.16 lb of N per palm a year; and the desert land nearby contained amounts of nitrogen varying at different depths and in different places from 23 to 630 lb of nitrogen an acre-foot; but, despite these supplies of nitrogen, the palms responded to additional nitrogen, either 40 lb of ammonium sulphate or 20 lb of that salt with 400 lb of dairy manure.

Nixon's advice (333) in 1959 for the Coachella Valley was that 1.81 to 2.72 kg of Nitrogen should be applied per year per palm, divided into two or three applications on sandy soil to reduce leaching.

These quantities should not be applied indiscriminately, however, since soils vary greatly in their requirements and it seems that sometimes no applications of nitrogen are necessary.

Furr _et al_ (179) concluded that the results of the manurial experiment that they carried out with date palms in the Coachella Valley, suggest that some of the soils there "of medium to heavy texture are well supplied with nitrogen and that long-continued, heavy annual applications of nitrogen on such soils are wasteful".

Furr and Barber (177) expressed the opinion that date growers in the Coachella Valley have been influenced in their manuring methods by the practice of the citrus growers and have applied more nitrogen than is necessary, that the excess has been lost and that the date palm, which grows all through the year and has a comparatively deep root system, probably can obtain all the nitrogen it needs from soil that contains less than is necessary for citrus.

Furr and Armstrong (173) report that the nitrogen content of the top 8 ft of date garden land and neighbouring virgin land in the Coachella Valley, varies from 2 000 lb an acre in some sands to 15 000 lb an acre in some loans, and that applications of nitrogen to the palms growing in sandy soils at the rate of 6 to 8 lb a palm resulted, with a time lag, in heavier yields of dates, but that in two trials in loam soil palms manured annually for five years with from 3 to 8 lb of nitrogen gave no greater yield than unmanured palms.

Monciero (287), from his experiments at El-Arfiane in the Oued-Rhir (Algeria), concluded that the date palm there needed no manuring because the irrigation water in the quantity applied provided for more than all their nutrient requirements including nitrogen. By proper irrigation and drainage a sufficient humidity is maintained in the soil so that the concentration of salts is enough for the requirements of the palms but does not reach the threshold of toxicity.

To sum up the general conclusion on the application of mineral fertilizers to dates is that they may be normally restricted to nitrogenous types and that the quantity needed, if any, must be established by trial and observation - with a keen eye on the "cost-benefit ratio".

As a general guide, it may be taken that two to three kg of nitrogen per annum per fully-developed palm, from all sources, is adequate in most soils (Arar, 21).

4. "Green" Manure and Cover Crops

The practice of ploughing-in a cover crop as green manure is in fact mostly limited to the U.S.A., but the growing of leguminous cover crops, especially lucerne (alfalfa) for fodder is common in young plantations in the Old World and must contribute to soil fertility through the crop residues.

When the soil is saline, barley is the best cover crop to grow between young date palms. Rice needs a large amount of irrigation water and its use is practicable only in exceptional circumstances, for example at Hofuf in Saudi Arabia and at Qurna in Iraq.

Barley withstands salinity better than alfalfa both as the germinating seed and when adult; but alfalfa can usually be sown after a year or so of leaching. The practice of growing vegetables under date palms, popular, for example, near Baghdad, around Cairo and in many cases, is discussed under associated crops in Chapter VI.

The cover crop most used in the Coachella Valley is Hubam clover, a variety of *Melilotus alba annua*, alternatives are sour clover and purple vetch.

CHAPTER VIII

IRRIGATION, SALINITY AND DRAINAGE

A. GENERAL

Irrigation is necessary for crops growing in places where the rainfall alone is insufficient. Such places are commonly those which are hot and where the rate of evaporation is high. The date palm, which needs moderate heat for vegetative growth and, while the fruit is ripening, great heat and no rain, is always an irrigated crop, except where its roots can reach an underground water table.

Irrigation is normally carried out by man; there are, however, several places such as the Wadi Du'an and the Wadi Hajar in the Hadhramaut where, although there is a little rain during the winter and none during the long summer, the magnificent date palms are not irrigated by hand. Rare spates come down these narrow wadis and their water is held up by weirs at frequent intervals. Each weir has its stone spillway but, as the spillways are placed alternately at the right hand and at the left hand end of the weir, the water takes a zigzag course, its velocity is progressively reduced and all, except what evaporates, sinks into the deep soil of the wadi bottom. Occasionally, there is a spate in the summer, when the influence of the southwest monsoon extends somewhat more westward than usual, but ordinarily there are only about half a dozen spates a year and those from autumn to spring. The deep soil of the narrow wadi with its rock bottom must act as a satisfactory reservoir for the water. It is possible also there are cross walls of underground rock which hold up the water.

In many places, the subsoil water is too saline for the palm, for example, on the northern shores of the southern Tunisian salt lakes, but, if the surface irrigation water is sweet and plentiful enough, the salt is kept down.

As the date palm is found so often in lands where soil and water salinity are high, the question of irrigation must always be considered in relation with the problems of salinity and the requirements of drainage.

Date palms in bed of the Wadi Hadhramaut irrigated only when the wadi comes down in spate.

B. IRRIGATION

1. Directed Rain Water

A date oasis may be at the foot of a mountain range. The winter rainfall is usually not enough to ripen the crop, but sufficient water is often obtained by constructing a dam along the floor of the valley parallel to the foot of the range, with an opening through which the accumulated run-off of rain water is led, in a specially constructed channel, to the date garden. If the soil and rock conditions under the garden are suitable, the water remains in the soil long enough to enable the palms to ripen a crop.

This system is widely used in southern Iran, for example, round Borazjan and on the island of Qishm. When the subsoil water in the garden falls too deep for the palms to reach it easily, they are often irrigated from an animal water hoist drawing from a lower level than that to which the palm roots have penetrated.

2. Flow Irrigation from Spring, River or Canal

The irrigation of a date garden may be direct from a spring or stream, or from a river, without any artificial aid to raise the level of the water, if the levels of water and land allow this. The water may flow out from the river or stream all through the year as, for example, when the garden is in a mountain defile and the stream descends rapidly, but most rivers in the date countries are subject to wide fluctuations. The Nile is the classic example of the summer rise, but although its various dams have reduced the amount of the fluctuations in its lower reaches, the palms on its banks still obtain their most copious irrigations in the summer. The Indus have similarly been largely controlled and also, more recently, the Tigris and Euphrates.

The only example of extensive tidal irrigation of date palms is on the banks of the Shatt-el-Arab and neighbouring Bahmanshir River. The flood tide backs up the fresh water in the river twice a day, so that this fresh water is forced into the creeks that take off at right angles from the banks at intervals of about one kilometre. At ebb tide the channels empty again. However, as the fresh water has become less in recent years and the salt water has increased, the date palms are being killed, those nearest the sea first, but the process continues further and further inland.

A good, straight, well-kept secondary channel in a Basra date-garden, Iraq.

The only other example of the tidal irrigation of date palms known to the writers is that on the banks of the Hindijan River in Iran, 150 km to the east of the Shatt-el-Arab, and also close to the Gulf.

If a date garden cannot be irrigated direct from a river or from a canal taken higher up its course, the canal being constructed so that its bed falls less rapidly than that of the river, then the level of water in the river can be raised by a weir or a dam so that the date land can be irrigated by flow from an off-take above that construction.

A particular type of flow irrigation is the kariz or qanaat, the underground channel, one metre, or a little more, high and less than a metre wide, generally excavated by small boys, and usually lined, except where it passes through rock, with bricks. Each brick is thin, elliptical and hollow, as it might be the somewhat flattened rim of a bicycle wheel. These elliptical rims are placed vertically one against the other inside the excavated channel, to which they make a lining, down which the water flows, and prevent a subsidence of the soil overhead from blocking the stream.

The head of the channel is at the source of water at the foot of a mountain range, 10 to 20 m or more below ground level. The maximum known depth of the first well of a qanaat, all dug out by hand, is said to be about 120 m. The land slopes down gently from the foot of the mountain to the valley bed, but the bed of the qanaat is excavated from the source of water horizontally so that it emerges at gound level somewhere between the mountain and the valley bed. Consequently, irrigation can be practised anywhere between this point of emergence and the valley bed.

At intervals between the source of water and the emergence of the underground channel at the surface of the slope between the mountain and the valley bed, and as the excavation of the channel progresses, wells are dug down to this artificial channel to allow the spoil to be hauled up. Round these wells the spoil is dumped, and form hillocks stretching across the dusty plains of southern Iran in all directions. The longest qanaats are over 30 km long. Iran is the home of the qanaat, but some can be found in Saudi Arabia, presumably dating from the time of the Persian occupation; and there are even a few, still in use, as far west as the Fezzan, wh ere they are known as fugaaraat (sing. fugaara), and where tradition ascribes their construction to the Romans.

Date palms killed by waterlogging caused by uncontrolled flow of new artesian wells. The ground is covered with rushes. Qatif, Saudi Arabia.

3. The Date Palm and Artesian Wells

Artesian wells are those drilled through an impervious stratum to release water which rises in the well because of the pressure to which it is subjected. Generally an artesian well is in a plain under which the run-off from neighbouring mountains has collected. For example, the artesian wells in the Ziban and Oued-Rhir of Algeria are supplied from the run-off of the Atlas Mountains, and those in the low-lying land of the Hasa Province of Saudi Arabia are believed to be supplied by the run-off from the mountains to the west of Arabia.

Because the supply of water from an artesian well is generally copious, constant and, once the original layout for drilling and lining the well has been incurred, free, there is widespread misuse of such water. The date gardens are over-irrigated and become salt marshes.

In Algeria in 1856, soon after the French assumed effective control of the Oued-Rhir, in that district there were 33 oases, all in a state of decay (Swingle, 419). The total flow of water from the springs was then 58 m^3 per min and the number of date palms was 136 000. Nineteen years later there were 22 natural springs, 492 flowing wells constructed by Algerians, and 114 flowing, tube wells constructed by the French. The total flow was now 254 m^3 a minute and the number of date palms of all ages was about 647 000 while the number of oases had been increased by ten. The population of the oases had more than doubled, and their value had increased more than fivefold.

Gautier (187) and Rème (378) describe the further development of the artesian resources of the Oued-Rhir. By 1890 the number of lined artesian wells had increased to 136 and the total output of water in the district was 309 m^3 per min. Since that year the output has hardly varied although the number of wells drilled had risen to 1 600 by 1933. In fact, the new wells had reduced the flow of the old ones. New gardens sprang up at the lower end of the valley and the older ones at the higher end died. As Rème (1935) pointed out, each new irrigated palm entailed the disappearance of an existing adult, so that the replacement of an old palm by a new one caused the loss of ten harvests!

Incidentally, Rème also states that the average amount of water received by the palm in the Oued-Rhir is only half of what it should be, 88 instead of 172 m^3 per year.

Precisely the same situation has arisen in the oasis of Qatif in Saudi Arabia, where the introduction ofthe art of drilling artesian wells by the neighbouring oil company in the 1930's has led to the drying up of springs on the high land, and to the ruin of gardens on the low land by over-irrigation.

In the date gardens of the Fezzan (Libya), the first artesian well was sunk by the Italian Government shortly before the Second World War, and after the war uncontrolled private entreprise has been responsible for sinking more. Here again, the flow of springs on the high ground has been reduced and useless lakes have been formed by the overflow of the new artesians.

In the neighbourhood of Riyadh, the capital of Saudi Arabia, the sinking of artesian wells to supply drinking water to the rapidly growing city has lowered the water table in the nearby date gardens, whose palms have died, except where their owner was rich enough to install a pumping plant.

Artesian wells, which are exceedingly useful if properly controlled, can do terrible damage if uncontrolled. No drilling of artesian wells should be sanctioned in places where date cultivation is important except under licence, to be granted only if the component Irrigation Department is satisfied that the date gardens will not be injured. Furthermore, the misuse of artesian water should be made an offence and rigorously suppressed. Records of output of water in date districts where there are artesian wells should be maintained by the Irrigation Department so that the effect of a new well on the old ones can be ascertained accurately and quickly.

4. Permanent Wilting Point and Irrigation Intervals

The date palm will flourish if it is provided with plenty of sweet water, and if the drainage is good. The amount of irrigation water it will stand is then very high.

The biggest palms in an oasis are those near the well or those growing in sandly soil close to a permanent water channel. However, water costs money and should not be applied in excess. It is not economical to supply the palm with more water than necessary for its transpiration, its food and its energy needs, or the soil with more water than enough to maintain its fertility. The moisture content of the soil should be maintained constantly above its permanent wilting percentage. The date palm, however, with its stiff, spear-like leaflets is not so quick to show if it is short of water as a soft-leaved tree, whose leaves wilt almost immediately the PWP in the soil is reached. It may be weeks, perhaps even months, before one can discover signs of drought on the date palm. Therefore, in the date garden, behaviour of other crops should be observed. When they wilt, the palms will also need water; but with practice one can establish the correct interval between waterings, which should be given just before the crops wilt. If there are no associated crops in the date orchard, broad-leaved weeds can be used as indicators.

As Nixon says (333) the frequency of irrigation will depend on soil texture and weather condition. Bearing gardens on the lighter soils in California are usually irrigated every seven to 14 days during mid-summer and every 20 to 30 days during winter. On heavier soils, irrigations are somewhat less frequent. In some localities where a permanent water table occurs at a depth of two to $2\frac{1}{2}$ metres, even as few as four to six irrigations a year appear to be adequate. In any case soils should be kept moist to a depth of two to $2\frac{1}{2}$ metres. The grower should make sure by means of a soil auger, tube or shovel, that the water is actually getting to the depths desired in all parts of the plantation.

With the date palm, which has no complete resting period (Mason, 271) it is possible that irrigation all the year round would be beneficial. It is commonly believed that the dates will cure more readily on the palm, if water is withheld when they are reaching the soft stage, but it may well be, that though drying may be accelerated by withholding water, the quality of the cured fruit may be inferior to that of palms whose irrigation was not interrupted. The competition of the fronds for water during the time of ripening of the fruit must not be forgotten.

However, it is usual to reduce or even stop irrigation during the harvest season. According to Arar (21) in his valuable paper to the Third FAO Date Conference, if ample soil moisture is provided up to early August for palms in full production on a deep soil of highwater holding capacity, subsequent irrigation may be greatly reduced or omitted entirely for two to three months or even longer, without reduction in yield or quality of fruit. Whatever irrigation programme is followed during the fall and winter, it is important that there be ample soil moisture available for palm growth during the spring and early summer, for a water deficiency then is likely to hasten the ripening of the fruit but also to reduce its size and quality. The amount of water used by crops in a given soil varies with the area of soil shaded by those crops, so that a young date orchard covered with alfalfa needs the same amount of water as one in which the palms are fully grown, with the tips of the fronds of one palm just not touching those of the next and with no cover crop.

Table 14

ESTIMATES OF VARIOUS AUTHORITIES OF THE WATER REQUIREMENTS OF THE DATE PALM

Date	Authority and Reference	Place	No. of palms/ha	M_3/ha/year	Gm/year	M_3/palm/year	Litres/palm/min	Remarks
1894	Rolland (287)	Sahara, Algeria	130	34 190	342	263	0.5	Monciero (1954, p. 36) mentions that this figure is still used by the Commission consultative des Forages dans l'Oued
1898	Rose (393)	Ziban, Algeria	144	10 368– 29 376	104 294	72 204	0.14 0.39	Annual average Summer average
1900	Jus	Oued-Rhir, Algeria	130	22 750 34 190	227 342	175 263	0.33 0.5	Minimum for existence For full-bearing palms
1935	Gautier (187)	Oued-Rhir, Algeria	129	33 927	339	263	0.5	
1935	Langronier	M'Zab, Algeria	130	17 940	179	138	0.26	Water sweeter than in Oued-Rhir
1935	Hannou (369)	Oued-Rhir, Algeria	129	38 700	387	300	0.57	
1935	Rème (378)	Oued-Rhir, Algeria	129	11 287 22 575	113 226	88 175	0.17 0.33	What the palms are actually getting What the palms should have
1955	Harris (194)	Qatif, Saudi Arabia	250	13 250	132	53	0.10	Not what the palms are actually getting but what he considers they need. His calculations are based on formulae used for various crops in the U.S.A.

Table 14 (continued)

Date	Authority and Reference	Place	No. of palms/ha	M_3/ha/year	Gm/year	M_3/palm/year	Litres/palm/min	Remarks
1957	Wertheimer (455)	Ziban, Algeria	120	15 000–18 000	150–180	125–150	0.23 0.29	
		Oued-Rhir	120	21 960–26 040	220–260	183–217	0.34 0.41	
1959	Nixon (333)	California	120	13 800–18 360	138–184	115–153	0.22 0.29	Heavy soil, annual rate
				18 360–27 480	184–275	153–229	0.29 0.44	Heavy soil, summer rate
				27 480–36 720	275–367	229–306	0.44 0.58	Light soil, annual rate
				36 720–55 080	367–551	306 459	0.58 0.87	Light soil, summer rate
1964	Stoler a/	Jordan Valley, Israel	120	25 550–32 850	255–328	213–274	0.41 0.52	
1970	UNESCO b/	Djerid, Tunisia	100	24 640–36 500		246–365	0.48 0.72	Quantities used successfully for Deglet Noor in project "Research and Training on Irrigation with Saline Water", TUN/5

a/ Personal communication to the authors

b/ Technical Report UNESCO/UNDP TUN/5, August 1970

5. Water Requirements of Date Palms

The principles governing the quantities of irrigation water which have to be applied are summarized by Arar as follows:

(i) compensation for evapo-transpiration of the palms with no stress;

(ii) compensation for evaporation losses from the soil surface;

(iii) provision for the leaching requirement so as to prevent the accumulation of salts in the root-zone;

(iv) provision of additional water required by inter-plantings or cover crops.

The first irrigation engineer to calculate the amount of irrigation water needed by the date palm appears to have been Rolland who, in 1894, put its needs at half a litre a minute throughout the year. This figure was still used in the Oued-Rhir of Algeria as recently as 1954 and, as it is calculated for full-bearing palms by dividing the output at source by the number of palms and, therefore, includes losses between source and palm when in a predominantly sandy soil, it would seem to be a good, average, basic figure.

Table 14 gives various estimates of water requirements for dates.

These figures are useful as a rough guide where new land is to be planted and irrigated by a source that is limited in amount and where the water table is low and cannot be tapped by the palms. The amount of water that a palm can obtain from the sub-soil water table may be deducted from its total requirements to give the amount that must be supplied from the surface; but this is true only if the sub-soil water is sweet. If the sub-soil water is salty, surface water will be necessary to dilute it sufficiently to make the mixture suitable for the palms and to prevent the accumulation of salt in the upper layers of the soil by capillarity. If the irrigation water is also somewhat saline, more will be required than if it were sweet. It is, therefore, only possible to calculate roughly the amount of irrigation water that will be required for any given new planting of date palms, but the more that is known about the soil and the two sources of water, the more accurate the estimate will be. A further complication is that in the date-growing countries, the salinity of both the irrigation water and sub-soil water may vary greatly with the season. For example, the water of the spring floods of the Tigris and Euphrates, which is used for winter cereals, used to have an Electrical Conductivity (E.C.) of about 0.016 mmhos/cm whereas the pumped water from the same rivers used for the irrigation of date palms and other crops in the late summer used to have an E.C. of 0.109 mmhos/cm. Sub-soil water in the date gardens bordering the Shatt-el-Arab varies in its E.C. between spring and autumn very widely.

6. Methods of Water Application

The following summary of irrigation methods and their application to date growing is taken from Abdullah Arar's paper (21).

(a) Furrow Irrigation

This method may be used for the different stages of the growth of the date palm, from the planting of the suckers to the fruiting stage. During the early stages of development one furrow running on one side of the line of the trees is normally used. An improvement is made by making two furrows, one on each side of the young trees.

In fully developed orchards, the area between the rows of the trees is divided by furrows spaced at 70 to 90 cm, and leaving a distance of about 30 cm in between the trunk of the tree and the nearest furrow. This method is to be preferred in heavy soils with low infiltration rate, so as to enable deep penetration of the water into the soil. The length of these furrows should not exceed 100 m in heavy soils in flat areas and less in sandy soils.

When land is not properly levelled before planting, the water flows rapidly to the lower end of the row and does not remain in contact with the upper and the middle parts of the row long enough for adequate penetration, especially when the slope is too steep. In such cases short furrows and a small stream of water in each furrow should be used.

(b) Basin Irrigation

Basin flooding in the areas of slow infiltration usually helps to overcome the difficulty of even distribution of water when layers of silt or silty clay exist in pockets which retard the downward water movement. It is also suitable for soils with high infiltration rate, where even distribution is attained by using small basins and high water discharges.

The basins can be very small, 2 x 2 m for the individual trees in the first few years, with a ditch running in the middle of two lines of the trees supplying water to these basins on both sides. The size of the basins could be enlarged with time until the sides of the neighbouring basins coincide. For fully developed orchards, large basins, one for four trees or more, could also be used, depending on the soil infiltration rate and discharge of irrigation water available. Low infiltration rate and high discharge of water calls for large basins and vice versa.

(c) Border Irrigation

Good land levelling is essential for this method of irrigation, more than is the case with furrow irrigation. Borders could be used for both young orchards and bearing palms. In the case of young orchards, the borders are made of strips about 2 m wide with earth dykes of about 30 cm high, with the row of the trees in the middle of the border. Lengths vary depending on the soil texture, slope and the amount of irrigation water available. The water is run in the border, while the spaces between the borders are not irrigated. Small circular ridges are made around each palm so as to avoid water coming into direct contact with the trunk. This system gives an economy of irrigation water while the date orchard is young. The width of the border (or strip) is increased with the increase of the age of the trees to about 3 m wide in the third year up to the sixth year. After this age it is replaced either by a complete border between the rows of the trees, as will be explained immediately, or by other suitable methods of irrigation.

The bearing palms are generally irrigated through border strips in between each two rows of palms. The level of the ridge on which the palms stand is about 30 cm higher than the border strip so as to avoid water touching the trunks. The length of the borders depends on the soil infiltration rate, land slope and the discharge of irrigation water available. High infiltration rate means short runs, slight slope short run, high water discharges, long run and vice versa. The optimum length of the border is arrived at after careful consideration of the combinations of these three parameters. Usually, with medium to heavy textured soil and with slight slope a border length of 100 to 200 m could be reached. In the case of light soils (sandy soils) a much lower figure is adopted and normally does not exceed 50 m.

(d) Sub-Surface Irrigation

This term might be applied to the method of irrigation being practised along the banks of the Shatt-el-Arab. This method utilizes the fresh water of the ebb and flow of the tide in a strip of land about 170 km long (from Kurnah, where the

Euphrates river joins the Tigris river to the FAO town on the Gulf) and a width varying from less than 1 km to more than 8 km on both sides of Shatt-el-Arab. It is based on the fact that the ebb and flow in the Gulf reverts the flow of the Shatt-el-Arab twice in 24 hours. Consequently, the irrigation network acts as drainage canals during the ebb and as irrigation canals during the flow period.

The network consists of creeks at right angles on both sides of the Shatt-el-Arab at a spcaing of about 1 km or more with a length varying frcm a few hundred metres to about 8 km.

From the creeks, water channels branch out also at right angles leading to irrigation ditches which form the main irrigation ditch for the individual orchards. The final ditches are those which are locally called Mekhlab, spaced at about 10 to 12 metres. The width of the Mekhlab is about one to three metres and the depth one to two metres. The trees are planted in rows on both banks of the Mekhlab. In this way, the date orchard is divided into snall pieces of land surrounded on three sides by irrigation ditches and each having two rows of date palms.

During the flow stage the level of water in the irrigation system is high, but normally is about 60 cm below the ground surface. Only in an exceptionally high flood season is the soil surface inundated. Consequently, the upper part of the soil profile is wetted by upward movement of water by capillarity.

In this way, the date palm is irrigated twice daily and the maximum level of water is usually reached at the beginning and the middle of the moon month. The normal variation in water level between ebb and flow is about one to two metres. However, from late summer to late winter the high water level in the canals drops to more than 1.0 m below normal which may result in orchards at the tail end of the canals not being properly irrigated.

This system of irrigation results in a relatively high water table, which will result in a shallow root system. Due also to the fact that the soil surface is not inundated, the salts that reach the root zone by capillary water in the upper 60 cm of the soil profile will not be leached back. This may result in a rather high salinity in this layer of the soil profile. Subsequently, the active root system of the date palms will be squeezed in between the upper salty layer of the soil profile (the top 60 cm) and the water table (1.0 to 1.5 m) below the ground surface.

Experiments have been made to improve this system in Basrah area, whereby surface water is provided by pumping and the water level in the old irrigation system is kept low by the use of check gates. Initial results indicate a pronounced difference between the growth of the date palms in favour of the surface method.

For this reason and the fact that the quality of the water of the Shatt-el-Arab will deteriorate in the future both in quality and quantity due to upstream irrigation development both in Iraq and Iran, the existing system of irrigation of the date palm of Shatt-el-Arab has to be changed into a surface syste. This is also justified by the fact that beside the problem of salinity and shallow rooted date palms mentioned above, the mechanization of the date palm cultivation, etc., is out of the question with the present system of irrigation. With rising labour wages one day mechanization will become "a must".

(e) Sprinkler Irrigation

This method of irrigation could be used with advantage in places where water is scarce and/or expensive. It is particularly justified in very light sandy soils where the infiltration rate is high and consequently the irrigation losses are excessive. It is also justified in heavy soils where topography is unsuitable for surface irrigation methods. This may be the case in very steep slopes or rough topography which will require a very high investment in land levelling. It has been reported by Reuveni (380), however, that the sodium content of the date palm leaflets has increased in orchards which were irrigated by sprinkler using a poor quality irrigation water. This problem is not considered serious as long as the irrigation water is not so saline that it could not be used for irrigating date palms anyway.

(f) Drip Irrigation

This new method of irrigation is considered the most efficient method from a water economy point of view. In places where irrigation water is expensive due to high head of pumping, this method might be economically feasible. Beside the high capital investment involved the problems of maintenance, including clogging of the drippers, should also be remembered. The bicarbonate and iron content of the irrigation water should be checked. As low as 1 ppm of iron or 100 to 200 ppm bicarbonate in the irrigation water can cause serious problems of clogging. However, under certain conditions, this system may deserve serious consideration for the date palm.

One of the advantages of drip irrigation is the ability to use irrigation water with a relatively high salt content, because the soil moisture in the root zone is kept close to the field capacity (Reuveni, 381, 382). In the U.A.E. forest plantations have been established with this system on deep, sandy soils using brackish water with salt contents as high as 7 000 ppm.

C. SALINITY

1. Alkaline and Saline Land

Alkaline land is that containing harmful quantities of alkalis containing the hydroxyl group - OH, especially caustic soda, NaOH. Saline land is that containing harmful quantities of the chlorides, sulphates and carbonates of sodium, potassium, magnesium and calcium. Alkali lands are among the most difficult ones to reclaim. Soils may be both saline and alkali.

In Iraq sone land is so saline that nothing grows. At the edges of such completely bare patches are found the plants that are the most resistant to salt - various species of *Suaeda*, *S. vermiculata*, Forsk, and *S. maritima*, Dum., and others, the commonest local names for which are *tahama*, *suwaad* and *suweida*. Another Chenopod, *Schanginia baccata* (Forsk) Moq., the *gagalla* of the Arabs, is usually found in rather less saline places. On somewhat sweeter land the halophitic grass, *Aeluropus villosus*, Trin., the *shuroib* or *acharish* of the Arabs is found, its leaves often sparkling in the sun with salt crystals. Then, where the crimson flowered camel thorn, *Alhagi maurorum*, Med., the *Aaguul* of the Arabs, begins, it is usually possible to grow an indifferent crop of barley. From where the common shok or kharnuum, *Prosopis stephaniana* (Willd.) Spreng., starts, anything can be grown. Dates, fruit, vegetables and alfalfa are normally planted in the last type of land because it is that of the high river banks, normally not flooded in spring, where drainage back into the rivers and inland to the lower, salt lands is good, and where water is available all the year round, in spring by flow and for the rest of the year by lift. See also Richards *et al* (384).

2. Cause of Saline Soils

For a general discussion of salinity problems, the reader is referred to Richards _et al_ (384) in USDA Handbook No. 60. These problems are typical of arid and semi-arid climates and, therefore, very familiar in date-growing countries. The major causes are that leaching and transport of soluble salts to the ocean is not so complete as in humid regions, and that the high evaporation rates tend further to concentrate the salts in soils and in surface water.

Frequently salinity is caused or increased in arid areas by irrigation when the drainage is inadequate for the quantities of water used.

The effect of the _quality_ of the irrigation water used is discussed in an FAO Bulletin on Water Quality for Agriculture (1955), which should be consulted for detailed information on the subject. Naturally, water with a relatively high content of dissolved salts is liable to cause accumulation of salts in the root zone and in some cases may also affect soil permeability. Good management may greatly mitigate these effects, however, as is indicated in the following paragraphs on leaching and drainage.

3. Tolerance of the Date Palm to Salinity

Richards _et al_ (381) give the following scale showing the relationship between crop responses and soil salinity expressed in terms of the conductivity of the saturation extract:

Table 15

RELATIONSHIP BETWEEN CROP RESPONSE AND SOIL SALINITY

(Richards *et al*, 1954)

Crop Response	Scale of Conductivity (Millimhos/cm at 25°C)
Salinity effects mostly negligible	0 - 2
Yields of very sensitive crops may be restricted (Radish 4*)	2 - 4
Yield of many crops restricted (Castor 6*)	
Only tolerant crops yield satisfactorily (Alfalfa c.9*) (Tomato 10*) (Garden beet 12*)	8 - 16
Only a few very tolerant crops yield satisfactorily (Barley 16*)	

* The electrical conductivity values of the saturation extract in millimhos per cm at 25° C associated with a 50 percent decrease in yield.

The date palm is shown alone in the "High Salt Tolerance" column in the table of fruit crops. The authors give medium salt tolerant fruit crops in this order: pomegranate, fig, olives, grape, cantaloup. Among the fruit crops of low salt tolerance, in decreasing order of tolerance, are mentioned apple, orange, grapefruit, almond and apricot.

The date palm is, in fact, more salt tolerant than any other fruit crop. It will survive in soils containing 3 percent soluble salts. About 6 percent of salt content of the soil is the limit above which date palms will not grow (21).

In the Arab literature, there are in fact occasional recommendations to growers to put salt in the hole in which date offshoots are to be planted and to sprinkle it round grown date palms, but from observations it would appear that the finest palms and the heaviest yields are found in sweet soil.

In this connection, it is of interest that Furr and Armstrong (175) carried out some experiments in the Coachella Valley in 1960 and 1961 on the effect of salt on the yield and growth of 17-year-old date palms in full-bearing. They added quantities of respectively, 100, 200 and 450 lb of sodium chloride on each experimental plot of two palms (3 450 sq. ft), applied to the surface before each watering from 19 May to 14 December 1960 and from 9 March to November 1961.

Their observations indicated that, in comparison with control plots, the variations in salinity apparently had little or no effect on growth rate of leaves, or yield, size, or quality of fruit, or on chloride content of the leaf pinnae. These results were remarkable, but perhaps final conclusions should not be drawn from them, since it is not known to what extent the salt had been washed down below the root zone.

4. Leaching, and the Use of Irrigation Water with Various Degrees of Salinity

Most of the salts added with the irrigation water are left behind in the soil, as water is removed by the crop. To dissolve and remove the salts, adequate water must be applied to allow percolation through the entire root zone.

The following Table 15 (Arar, 21) shows the crop tolerance and leaching requirement of some important crops, including date palms. Expected decreases in yield from zero to 50 percent are indicated, with the use of different qualities of irrigation water for each crop, with the relevant leaching percentage required. It will be noted that it is possible to irrigate date palms with water of salinity up to 3.5 mmhos/cm (i.e., about 2240 ppm) with no reduction in yield, provided that a leaching requirement of 7 percent is provided for. With irrigation water of 5.3 mmhos/cm salt content and with leaching requirement of 11 percent, the reduction in yield is only 10 percent. However, 50 percent reduction in yield is expected when the salt content of the irrigation water is as high as 10 mmhos/cm (about 6400 ppm) and with a leaching requirement of 21 percent. These figures demonstrate how very salt tolerant the date palms are. From the same table it could also be noted that barley, sugar beet and cotton are rather salt tolerant, while citrus and strawberry are very sensitive to salinity.

Table 16

CROP TOLERANCE AND LEACHING REQUIREMENT

Yield Decrement to be Expected for Certain Crops due to Salinity of Irrigation Water when Common Surface Irrigation Methods are Used

Crop	0 %			10 %			25 %			50 %			Maximum
	ECe 1/	ECw 2/	LR 3/	ECe	ECw	LR	ECe	ECw	LR	ECe	ECw	LR	ECdw 4/
1. Barley	8	5.3	12 %	12	8	18 %	16	10.7	24 %	18	12	27 %	44
2. Sugar Beet	6.7	4.5	11 %	10	6.7	16 %	13	8.7	21 %	16	10.7	26 %	42
3. Cotton	6.7	4.5	11 %	10	6.7	16 %	12	8	19 %	16	10.7	26 %	42
4. Wheat	4.7	3.1	8 %	7	4.7	12 %	10	6.7	17 %	14	9.3	23 %	40
5. Rice	3.3	2.2	12 %	5	3.3	18 %	6	4	22 %	7	4.7	26 %	18
6. Beans	1	0.7	6 %	1.5	1	8 %	2	1.3	11 %	3.5	2.3	19 %	12
7. Figs, Olives	3.3	2	8 %	5	3.5	12 %	-	-	-	9	6	21 %	28
8. Citrus	1.7	1.1	7 %	2.5	1.7	11 %	-	-	-	5	3.3	33 %	16
9. Strawberries	1.0	0.7	7 %	1.5	1.0	10 %	-	-	-	3	2	20 %	10
10. Date Palm	5.3	3.5	7 %	8	5.3	11 %	-	-	-	16	10	21 %	48

1/ ECe means Electrical Conductivity of soil saturation extract in milliohms per centimeter (mmohs/cm)

2/ ECw means Electrical Conductivity of irrigation water in mmhos/cm

3/ LR means Leaching Requirement

4/ ECdw means maximum concentration of salts that can occur in drainage water under crops due to ET

N.B.: For conversion to TDS as PPM multiply mmhos/cm by 640.

The main drain at Saba Abyaar, near Tozeur, Tunisia, being straightened and deepened.

The same drain a little higher up, where the required depth has almost been reached. Saba Abyaar, Tunisia.

It is possible to apply the additional quantity of water required for leaching at every irrigation and under these conditions the soil, if it is sufficiently drained, is not in danger of salt accumulation.

However, new research information (155) indicates that this method of leaching may not always be necessary and that salt accumulation can take place for short periods of time in the lower root zone without adverse effects. The roots of the date plam usually extend to a depth of up to 2 m, but 50 percent of the water needs of the palm are absorbed in the upper 60 cm of the root zone and 80 percent in the upper 120 cm (21). As more water percolates through the upper root zone than the lower, the upper root zone will contain less salinity. It is, therefore, possible to rely, to some extent, on seasonal leaching, provided that the salts in the more critical upper root zone are not allowed to reach damaging concentrations. This is an especially important consideration in date growing regions, such as those in Tunisia and Algeria, where there is a fairly copious winter rainfall. Naturally, the timing of the leaching must be adjusted in each case, according to the qualities of soil and water, the conditions of drainage, and the characteristics of rainfall.

Valuable information on practical management in these matters may be obtained from the reports of the UNESCO/UNDP project "Research and Training on Irrigation with Saline Water" (TUN/5), which was carried out in several stations in Tunisia from 1962 to 1969, and of the follow-up FAO/UNDP project (TUN/29), as well as from the FAO Bulletin on Water Quality (155).

D. DRAINAGE

It would not be suitable to go into full details of the calculation of drainage requirements in a study of this type, but some of the principles involved are mentioned here. Agronomists concerned with the design or management of date plantations would be well advised to consult a drainage engineer to ensure that adequate provisions are made, since drainage problems are the commonest cause of poor growth and yield in most date growing countries.

It should be remembered that although date palms are resistant to occasional flooding for some days or even weeks (21), they used a well-drained soil for healthy growth. "When bringing new lands under irrigation, planners have frequently failed to recognize the need for establishing artificial drains to cope with the additional water and soluble salts. As a result, the ground water table may rise from a considerable depth to within a few feet of the soil surface in a few years. During the early development of irrigation projects, water is frequently plentiful and there is a tendency to use it in excess. This hastens the rise of the water table. When the water table rises to within 1½ to 2 m of the soil surface, ground water moves upward into the root zone and to the soil surface. Under such conditions, ground waters as well as irrigation water contribute to the salinization of the soil" (21). It is clear that drainage must always go hand in hand with irrigation development.

The calculations for the lay-out, type and depth of the drains will be based on the topography of the site, the type of soil, the irrigation and leaching requirements, and the rainfall, and should have as the main objective keeping the water table below a depth of 150 cm most of the time (21). "Covered field drains should be used whenever possible in view of their low cost of construction, if carried out with modern trenching machines", their low maintenance cost, and because they leave the date garden suitably free for irrigation, cultivation and harvesting operations.

CHAPTER IX

CULTURAL PRACTICES 1/

A. TILLAGE

On the basis of the Shatt-el-Arab the ideal that the rich cultivator used to follow and the poor one tried to follow was a deep digging every four years and a shallow one each of the other three. The appearance of the palms in gardens receiving this treatment compared with that of the palms in gardens with little or no digging, showed that such cultivation was beneficial. The circumstances, of course, were peculiar: the soil was an extremely fine, tenacious clay which, when soaked with the spring floods and baked by the fierce summer sun, dried into a mass of almost brick-like consistency, into which date palm roots appeared to penetrate only with difficulty. Rising labour costs have resulted in a great reduction in the amount of digging on the banks of the Shatt-el-Arab and in consequence, in the vigour of the palms and in the quality of the dates.

Probably the results of this laborious cultivation were also due to the fact that, under the system of irrigation used, water did not reach the surface. Unless the manure was incorporated in the soil below a depth of 50 cm or so, it was never adequately moistened and the palms did not obtain the extra nitrogen they needed for good growth and cropping.

Deep cultivation of any sort around palms presumably also has its negative aspects, as it does with other tree crops, because of the damage it does to the root system. When surface systems of irrigation are used it would, therefore, seem best to limit tillage to shallow harrowing to keep the surface clean and level and to incorporate applications of fertilizers. Ploughing will be needed when growing intercrops, but the latter will usually only be important when the palms are young or when the rows are very widely spaced. One way or another, however, the farmer will always need some means of cultivation and the instrument he uses will depend primarily on the economics of his business and the size of his farm.

1/ For notes on pollination and harvesting, the reader is referred to Chaps. V and X.

Basra, Iraq.
The first digging (or Kusuur) is done by men in groups of three. After the kusuur comes the nashuur, the second digging, which is not such hard work and is done by men working singly in line.

With the rising cost of labour resulting from the development of the oil and other industries in the date-producing countries and from the improvement in means of communication, which makes it easy for the young cultivator to leave the land for the town, the date-garden owner may be expected to substitute machinery for hand labour. Raising irrigation water by motor pump is now generally cheaper than raising it by hand, but tractor tillage, although used in all the date lands of California, has made little progress in the Old World. Here, both the yield and the price of dates are lower than in America, and the cost of labour also, though rising, is still lower than across the Atlantic; but perhaps the chief difficulty in using tractors in the Old World is the small size of the individual date gardens. A tractor does not pay for itself unless it is kept busy, but if a man has only a small garden, he cannot keep it so. The difficulty of using machinery in small gardens to some extent has been overcome on the coast of Libya by the rise of a class of entrepreneurs who own tractors and who cultivate the land with them at so much per hectare. Such tractors were working in the Zliten district, where the palms are planted in widely separated rows.

For smallholdings, the two-wheeled horticultural tractor or rototiller or the miniature track-laying tractor may be found suitable, although many peasant date growers will doubtless continue to use their hoes and their light animal-drawn ploughs. Nevertheless, as a general rule, when villagers see a clear advantage in modern equipment and when it is offered to them on long-term payment, they will accept it. The very wide adoption of small oil-engine water pumps in recent years is a good illustration of this.

B. WEED CONTROL

Weeds compete with the palms for water and soil nutrients and, unless leguminous, add to the soil only humus, but no nitrogen, when they are ploughed in. Where the air is humid at harvest time, as for example in the Coachella Valley of California, it is good practice to plough weeds in before harvest so that they do not contribute to the humidity of the air in the garden, otherwise annual weeds usually do but little damage in date gardens. There are a few species, however, which can be very troublesome.

The common rust (*Juncus*, probably *maritimus* L.) will sometimes smother young palms growing in swampy places, for example, at Taworgha in Libya and Qatif

The rototiller is a useful implement in land much cut up by channels. Basra, Iraq.

in Saudi Arabia, but the remedy is, of course, to drain the garden. The drainage will kill the rust and improve the palms.

Bermuda Grass (Cynodon dactylon Pers.), Devil-, Dog's Tooth-, or Wire-Grass; in Arabic Irg en Najiil (ARE), Kazmiir, Najiil, Najiir, Najm, Najma, Najm es Saliib, Tahma, Tahmaa, Theiyil (Iraq), or Ubeil (Yemen), can be a bad weed in young date gardens, but in gardens of adult palms, is usually considered an asset by the cultivator because it provides him with fodder for his cattle and does not make a dense growth in the semi-shade of the palms.

The worst weed in date gardens is probably Imperata Grass (Imperata cylindrica, (1) P. Beauv., the Lalang of Malaya, the Abu Duwois or Sill of the Arab Republic of Egypt, the Halfa of Iraq and the Diis of Algeria.

Uncontrolled, this grass grows from one metre to one metre and a half high. It has very strong, stiff stems, armed with a sharp point; and it is hard and painful to push through it. It will smother young palms, and cause tall ones to loose their vigour. Its roots will bore through the roots of date palms and kill them and its shoots will grow up a date palm stem under its frond bases. At Tolga in Algeria, palms have been found with three-quarters of their roots killed by Imperata. The same trouble has been reported from Tiberias and north of Eilath in Israel. Burkill states that burning is no use as a control measure. Helgeson (200) states that "sodium arsenite has been used in young rubber plantings. A 2.5 percent solution has been employed but a 1 percent solution is sufficient if wetting agents are used. Apply with stirrup or bucket pumps at the rate of 60 to 80 gallons per acre. Four or five sprayings are required for control. After the initial sprayings with the arsenical it has been the practice to wipe the new shoots with oily rags at four-week intervals". It would seem that spraying with a knapsack sprayer using a fortified oil emulsion would probably be just as effective and perhaps less expensive. Wiping new shoots with oily rags once a month would seem to be a formidable task.

Some of the modern weed killers are powerful and persistent and should not be used in date gardens where drainage water is used as irrigation water by other date gardens at a lower level, if there is a likelihood of those weed killers getting into the drainage water of the upper gardens. See Helgeson (200) for precautions to be taken when weed killers are used in orchards. "Dalapon" and

Halfa (Imperata cylindrica, the lalang of Malaysia) in a date garden. It will smother and kill small date palms and reduce the yield of adults. Khuzistan, Iran.

Plexus of halfa roots and stolons from a little below the surface of the soil. Khuzistan, Iran.

<u>Imperata</u> grass (Halfa) flowering and seeding. Ahwaz, Iran.

"Dowpon" have been recommended. These products were widely used in 1966 on Lalang in the rubber plantations of Malaysia and give excellent control. For many years now citrus growers have kept down weeds with weed killers and have found that cultivation is then unnecessary; and this procedure for date gardens has been followed more recently in California and Israel with satisfactory results.

C. PRUNING

What is often called pruning in respect of the date palm is not the same as the pruning that the fruit trees and bushes of temperate fruits receive. These have living wood removed for the purpose of thinning or shaping the growth.

"Pruning" of the date palm is usually only the removal of dead, or nearly dead wood, in fact the dying fronds and, when their bases have also dried out, of these also.

The exception to the above statement is the occasional removal of green leaves from Deglet Noor palms in California. About this, Nixon (333) writes as follows: "As many as 180 green leaves have been counted on a single palm unpruned for six years. When this many leaves are retained, a considerable number will be below the fruit bunches, and in certain localities they may increase the percentage of fruit affected with checking and blacknose and of fruit in the dry grades. These excess lower leaves apparently increase the relative humidity around the fruit bunches in mid-summer when checking occurs. Later, when the harvest begins, these leaves probably complete with the fruit or water, especially under the usual practice of reducing the amount of water applied during harvest. Under such conditions it is desirable to remove some of the lower leaves in order to improve the fruit quality. With the Deglet Noor, which has long fruit stalks, removing the leaves up to about the point where the lower ends of most fruit bunches are exposed has proved satisfactory on palms in full bearing".

These remarks, however, refer only to young Deglet Noor palms and only to California. Nixon goes on to explain that when the Deglet Noor are over 20 years old they seldom retain leaves below the mature fruit bunches. With varieties that have short fruit stalks like Halawy and Khadrawy, removing green leaves below the fruit bunches will reduce the bearing capacity of the palm.

A nice example of the frond-base cutter's art. A Khadhraawi palm at Abadan, Iran.

It may be considered an advantage to remove spines from the new leaves every year, so as to facilitate pollination and handling of fruit bunches. In America this is done with a long curved pruning knife mounted on a long handle.

D. FRUIT THINNING

Two main types of fruit thinning may be practised - removal of whole bunches and removal of a proportion of the fruit from each bunch.

1. Bunch Removal

The type of thinning used both in the New and Old World date growing regions is the removal of entire bunches. This was the regular practice in the well kept gardens in Iraq and Algeria, although it has recently become less common. It consists in cutting off some of the bunches when it is considered that their number is so large that, if not reduced, the palm will bear little the following year. Dowson remarks that, according to the size of the palm and the fancy of the garden owner, the Basra practice was to leave on the Hallaawi palms 8, 10, 12 or 14 bunches but not more than 14. The bunches cut off were fed to the cattle.

In California also, the safe load (333) for Deglet Noor is considered to be 12 to 15 bunches, with perhaps two or three more if the palm is exceptionally vigorous.

Nixon comments that Deglet Noor palms in full production and pruned to 100 to 120 leaves have been able to carry one moderately thinned bunch of fruit for each eight or nine leaves without reducing the number of flower clusters in the following year. With this variety it is common practice to remove the earliest and latest bunches which are often smaller than the others.

Young palms should be allowed to carry fewer bunches. None for the first three years; one or two in the fourth year; three or four in the fifth year and so on. Production should increase yearly until the maximum number and size of leaves are reached - usually in 10 to 15 years depending on the variety and growing conditions.

2. Fruit Thinning in the Bunches

Nixon gives a detailed account of the California practice of fruit thinning in the bunch. Whether this technique should be adopted in Old World date growing countries depends partly on climate and partly on market preferences. So far it has not usually been considered necessary.

The relationship to climate is important, as part of the object of the exercise in California is to allow a greater circulation of air round the fruit of the Deglet Noor and thus reduce the damage due to humidity. In other climates and with other varieties this consideration would not be so important. Market preferences may make fruit thinning advantageous if, for any particular variety, it is required to increase the size of the fruit. However, Al Bakir and Al Azzawi (7) after four years trials of various kinds of thinning of Zahidi dates in Baghdad, reported that no thinning was more profitable than any of the five types of thinning tried.

If this type of thinning is to be practised at all, early thinning is better than late thinning. The former allows room for the development of the dates that are to remain on the bunch from an earlier stage, requires less labour and there is less loss of the elements that have commonly to be replaced by manuring, namely nitrogen, phosphorus and potassium.

There is, however, another side to the question. If the ravages of the lesser Date Moth are serious, the fruit may be damaged after 10 June, so that by thinning before that date too many dates may be removed. Similarly, if locusts are likely to visit the groves it may be wise to postpone thinning until after the visit when it may be found unnecessary. Also, the larger the dates are when thinned, the greater is the amount of food for the cattle. This last consideration is of no interest to Californian growers, but is a matter of some importance in countries, Baluchistan for example, where pasture for the sheep, goats, donkeys, cows and camels is scanty and where the bountiful date palm is almost the universal provider.

The technique of fruit thinning of Deglet Noor as described by Nixon is as follows:

(a) the tips of all strands are cut back enough to remove one-third or slightly more of the total numbers of flowers or fruits;

(b) the central strands are removed so as to reduce the total number of strands in the bunch by one-third to one-half.

With other varieties of dates, the technique is modified as to the amount, if any, of tipping the strands and the number of strands removed. Reducing the number of fruits per bunch increases the size of fruit but cutting back strands may increase the susceptibility of fruit to "checking" - fine cracking of the fruit skin or "blacknose" - a darkening and shrivelling of the tip.

The final result should be that each bunch will carry 20 to 35 dates per strand and 30 to 50 strands.

E. PROTECTION FROM RAIN, INSECTS, BIRDS AND DRY WIND

In California and Arizona bunch covers of paper or other materials were devised to protect the dates from the rain. This function they performed satisfactorily, but they reduced the rate of circulation of the air round the dates on the bunch so that, although the dates escaped damage by actual wetting, damage by excessive humidity increased. It was found, however, if in addition to reducing the number, or the length, or both of the spikelets, the remaining spikelets were separated from each other, damage by humidity was reduced. To effect this separation starshaped wire rings were inserted in the centres of the bunches (Dillman et al). The paper bags or tubes are attached to the fruit stalks immediately above the bunches with the lower ends left open. They are usually put on after the fruit begins to acquire its khalal colour. (N.B. Boiling the khalaal dates in the hard, yellow stage to make "khalaal matbuukh" is a good way of saving rain-damaged dates, if there is a market for such cooked dates).

In Israel, Hilgeman (201) observed such serious damage to the ripening dates by wasps that he considered them to be the worst pest of dates there. This pest now does no more damage, because all bunches are enclosed in half-metre square bags of wire mesh. Birds, of course, are also prevented from damaging the fruit. Each bag costs about US$1.25 and lasts for five or six seasons.

In the plantations to the north of the Gulf of Akaba, where the air is too dry for the best Deglet Noor production and where there is a high proportion of dry dates, quality has been improved by enclosing the bunches in plastic bags before the dates start to soften.

Dalaawa, that is, the bending down of the bunches whose dates are in the green stage, so that they rest on the lower fronds. This practice prevents or reduces the breaking of the bunch main axis, and facilitates harvesting. Ahwaz, Iran.

F. SEPARATING BUNCHES

Some varieties, for example the Deglet Noor and the Barhi, have very long bunch main axes. Their bunches, therefore, tend to remain separate; but there are some varieties, such as the Kungard of Bandar Abbas, which have very short main axes. The bunches of most varieties are between these two extremes. Generally, it is desirable for the bunches which have intermingled to be separated at an early stage in the development of the dates, that is to say when they are still green and growing out of the spherical stage and becoming long. When they are like this, the dates are firmly attached to the spikelets. At a later stage, there is likely to be loss owing to the dates of adjacent bunches being more tightly intermingled and being less strongly attached to the spikelets and, therefore, falling to the ground more readily when an attempt is made to disentanble them. This operation in Arabic is called "tafrid", i.e. "separation".

Lowering Bunches

When the dates are almost fully grown, but still green, to reduce breakage of their main axes and to facilitate harvesting, it is customary to pull them downwards till they come to rest on the lower fronds of the palm. With the very long stalked Deglet Noor in North Africa and California the bunches are brought down even lower, to below the bottom fronds, to which their main axes are tied with twisted frond leaflets or with rope or twine.

Supporting Bunches

As mentioned above, the bunches may be supported by the lower fronds, but if these are not strong enough to take all the weight, various devices, such as crutched sticks stuck in the ground or against the trunk of the palm, are used.

G. TIMING OF FIELD OPERATIONS

To conclude this chapter an extract is given in Table 17 of some old records of a large plantation in Basra, showing the development of the crop and the progress of operations from the first appearance of the male spathes in January to the end of the harvest in September.

On low palms supports of dried fronds stuck in the ground assist the lower, living fronds to take the weight of the dates. The variety shown is Barban, a heavy yielder. Ahwaz, Iran.

Table 17

DAY OF MONTH AND MONTH OF POLLINATION, HARVESTING, ETC.

FROM THE RECORDS OF A LARGE DATE PLANTATION IN BASRA

Year	Male Spathe		Female Spathe		Pollination		Lowering Bunches		Mite			Rutab	Bureim Harvest		Hallaawi Harvest		No of days from pollination (begun) to Hallaawi harvest (begun)	Remarks
	1st appeared	1st split	1st appeared	1st split	Begun	Ended	Begun	Ended (d)	Web 1st seen	Sulphring begun	Sulphuring ended	1st seen	Begun	Ended	Begun	Ended (all varieties)		
1924					10.2	7.4							18.8	22.9	21.8		164	
1925					31.3	13.4							10.8	22.9	28.8		150	
1926					15.3	28.3							13.8	27.9	21.8		162	
1927					21.3	25.4							22.8	10.10	11.9		175	
1928				3.3	19.3								16.8	24.9	19.8	8.9	153	
1929	15.2		30.12.28		27.2	10.4						10.7		6.9*	16.8	28.8	170	
1930	3.1		15.1		22.2										20.8		181	
1931	8.1		7.1		23.2	15.4									15.8		173	
1932	3.2	14.2	3.2	3.3	12.3	24.4	4.6			8.6	6.7+		10.8		13.8		162	Shaking Hallaawi begun 13.8, cutting 20.8
1933	13.2			13.2	20.3	7.5	27.5	30.6	4.6	1.6	3.8		21.9 (a) 19.8 (b)		2.9	30.9	166	Web seen only twice on Estate Bad in neighbouring gardens
1934	10.2		27.12.33	25.2	5.3	30.4		7.8	17.6	18.6	27.8			29.9	8.8	27.9	156	Shaking Hallaawi began 8.8, cutting 26.8
1938					19.3	8.5		19.7	12.6	15.6	31.8							
1939			4.2		8.3	4.5	4.6	22.6	6.6	6.6	17.8	29.7	17.8		26.8		162	Harvest of Bureim begun 17.8
Average	29.1	14.2	14.1	15.2	11.3	20.4	2.6	26.6 (e)	9.6	9.6	16.8 (f)	19.7	16.8	24.9	22.8	13.9	164	

NOTES:
- * All except Liilwi
- a. Ista 'amraan
- b. Bureim
- c. Except for Liilwi and Khasaab
- d. The day and month in the column "Lowering Bunches ended" for the years 1934 and 1938 does not refer to this operation but to the second operation of placing the heavy bunches on the lower fronds of the palm for support
- e. 1933 and 1939 only
- f. 1932 excluded

CHAPTER X

HARVESTING

The way in which the dates are harvested affects the packer directly. If they are harvested in a clean way, the packer is saved the labour of cleaning, while if they are allowed to fall on the ground and pick up soil and sand, the packer is given the extra work of removing the added dirt. Furthermore, the time and method of harvesting and the care given to it largely determine what processing is required.

One of the difficulties with which modern date-packing houses in various countries have to contend, and one which adds considerably to their operating costs, is the poor state in which the dates arrive at their doors, due to improper harvesting methods. The aim of agricultural extension services in date-growing districts should, therefore, be to encourage the use of garden boxes and the placing of dates directly from the palm into the box.

REACHING THE DATES

There is no difficulty in reaching the dates on low palms, but most date palms are tall and have to be climbed.

Palms may be climbed barefoot, which is the usual way, but it is tiring, risky and makes carrying implements up the palm and their use at the top difficult. Footholds may be scooped out of the trunk of a tall palm to assist climbing (Figure 13).

Sometimes a man will climb a palm unaided, carrying a rope with him with which he will support himself when he reaches the top. Sometimes he will also use a simple rope belt while he climbs (Figure 14).

Climbing a date palm with the aid of a climbing belt. Iraq.

An improvement on the simple length of rope is the wasla of Libya (literally, that which causes one to arrive), which consists of a stout rope to encircle the palm, and a large number of small ropes similar to a saddle girth for supporting the climber's back. This belt is also used in Algeria.

An improvement on the wasla of Libya is the farwand of Iraq (perhaps from the Persian word parwandeh, meaning rope). Farward is also the name in Khuzistan (Iran), parwan in Bushire (Iran), and parbun in Panjgur (Makran, Pakistan). In Kech (Makran, Pakistan), however, the belt is called tor, in Bahrein karr, in the Yemen markadh. The Iraqi farwand nowadays has a wire rope round the palm and a wide belt woven out of coir. The end of the wire is tied to a wooden peg which is inserted in the rope loop.

It is generally the men who climb the date palms, but the women, as well as men and children, collect the dates from the ground. At Darj, near Ghadames in Libya, however, the women of the negro cultivators climd the palms as readily as the men.

In the United States, the date farmers are reluctant to remove their shoes and socks and to climb up barefoot. Instead they employ mechanical devices.

Date growers in the United States are constantly changing and improving their methods of harvesting dates. Early date growers once used three-legged step ladders when the palms were young. They now use short light aluminium ladders for small palms. Wooden picking platforms suspended on chaims from the crown of the tree with attached ladder were tried but found too cumbersome to raise as the gree grew taller. Another innovation was an extendible ladder on flanged wheels which ran on a circular track around the palm. A chain from the top of the ladder semicircled the palm and prevented the ladder, which was tilted away from the palm, from falling. This was, however, found to be too heavy and moving it from tree to tree too time consuming.

Extendible towers were developed during the Second World War with two-wheeled and four-wheeled models drawn by a tractor. On the four-wheeled towers, four palms could be worked simultaneously from eight adjustable catwalks which extended out two for each tree. Sixteen men could be employed, but this was not efficient as the tower advanced only as fast as the slowest man. Forty to 50 ft aluminium extendible ladders proved more adaptable as they could be handled by one man (Figure 16). The picking saddle, similar to a Bosun's chair is used in conjunction with the aluminium extendible ladder. The date picker buckles on the saddle, throws the hook and safety chain over his shoulder, climbs the aluminium ladder until he can reach the date bunches, throws the hook over the base of a strong frond, adjusts the safety chain, puts his feet against the tree, and thus supported in the saddle, picks the dates into a bucket snapped to the saddle. When the bucket is full he lowers it to the ground on a rope, unhooks it, hooks on an empty bucket which he raises into the tree and moves around the tree to the next bunch.

The latest development is the hydraulic boom mounted on a truck with a basket mounted on the end of the boom. A man stands in the metal basket and is lifted to the necessary height for performing the harvesting operations, which consist of removing the entire date bunch and placing it in the basket. When all the bunches have been cut from the tree and placed in the basket, the boom is lowered and the date bunches are released into the shaker, which removes the dates from the bunch in as little as three and one half seconds per bunch, dropping them into a bulk bin which can contain 800 to 1 000 lb.

DETACHING THE DATES

For household use, the owner will pick the rutab dates one by one, and in the United States some growers still pick rutab Deglet Noor in this way. Dates picked one by one and placed in a receptacle do not become squashed or dirty, but such picking is slow and laborious.

Dates seldom ripen evenly on the bunch, though the hotter the climate, the shorter the time between the ripening of the first date and that of the last one on the same palm. In marginal lands, as much as two months may elapse between the appearance of the first rutab date and that of the last one on the same palm. In these places frequent hand picking ensures almost uniform maturity in each consignment received by the packing house.

Shaking the bunch by hand will cause the rutab and tamar to fall while the khalaal will remain attached to the spikelets, that is, to the stalks. If a receptacle is placed under the bunch, the dates will be caught and will not become dirty.

At Tump and Panjgur in Makran, the rutab are either shaken into a basket or into a piece of cloth tied to the cultivator's waist. The opposite edge of the cloth is stitched to a stick, to the middle of which a hook is attached. The cloth is stretched out under a bunch and the hook is slipped over a frond on the side away from the cultivator. This method is also used in the United States with soft varieties such as Khadhraawi.

In Makran, also to prevent dates falling from the bunch to the ground and being spoiled or lost, a jar-shaped basket called sund, woven from leaflets of the dwarf palm, is put over the whole bunch and a string, made of leaflets of the same palm, pulls the neck tightly round the main axis of the bunch. When the whole bunch is ripe, it is cut off together with the basket.

Lowering a bunch by rope. Iran.

Cutting off the whole bunch is the harvesting method used for the greater part of the date crop (Figure 18).

LOWERING THE DATES TO THE GROUND

If the dates are tamar and, therefore, moderately hard and not sticky, the ordinary way of getting them from the top of the palm to the ground is by throwing the whole severed bunch down. Mats or sacking may, or may not, be placed on the ground.

In parts of the Sahara, in Tunisia, and in Ghadames in Libya, a chain of men, from the crown of the palm to the ground, passes each bunch down by hand. The men are usually cultivators of neighbouring plots who, after finishing one plot, go in a group to the next one, until all the dates belonging to the whole group are harvested. No implements are required, except the sickle, and the dates reach the ground clean and undamaged.

In Mexico, the bunches of Khalaal may be slid down a rope.

In Iraq, usually, and in Libya often, the bunches of khalaal and rutab dates are lowered on a crutched stick (maglas, that which is attached, in Iraq; makhtaf, hook, in Libya) tied to a rope (Figure). A man on the ground removes the bunch and places it on a mat and the harvester hauls the crutch up again by the rope. It has been reported that the crutched stick has been modified in eastern Baghdad (Baquba area) to a lever pulley type that does not have to go down with the rope. Only the bunch goes down and the pulley remains temporarily tied to a leaf base. It is a convenient piece of equipment that can still further be improved to make the lowering of date bunches an easy operation.

COLLECTING THE DATES FROM THE GROUND

When the bunch is thrown to the ground, most of the dates fall off through the force of the impact. Those that remain are removed by shaking the bunch or, if that is not effective, by picking them off the spikelets one by one.

In the Fezzan, the dry dates are thrown directly onto the sand. In Iraq, the bunches are usually thrown onto small mats woven from date leaflets, but many of the dates fall outside or bounce off the mat.

An improved mat was adopted in a big plantation in Basra. It was larger than the ordinary one and wider at one end than the other. The two side edges were sown over to make hems into which two short lengths of light bambos were inserted. These provided an easy means of tipping the dates off the mat at the narrower end into the field boxes. The mats being narrower at one end than at the other when placed side by side, made sectors of an annulus and could thus be set round the trunk of the palm. If mats are not available, old sacks or sacking are often used.

When the main crop of dates is picked up from the ground or from a mat, the windfalls are commonly collected also and mixed with them. As the windfalls are generally infested and the cut dates are not, it would be better to keep the two lots separate. This can be done when the grower is his own packer. Otherwise the grower has no incentive to keep them separate, unless an infestation count is made of each parcel at the packing house and the purchase price made to vary inversely with the degree of infestation. This is a recommended practice for packing houses.

FIELD CONTAINERS

The field container in many Old World countries is a basket made out of a long, narrow strip of woven date palm leaflet matting sewn onto itself spirally. Handles of date palm fiber are sewn into the rim of the basket on either side. Sometimes baskets made of esparto grass are joined together to make panniers.

If the dates are hard and dry, they are generally put straight into sacks either of hard-wearing wool that lasts for years, or nowadays, often into gunny bags.

In the United Arab Republic, where in Cairo and the other large towns there is a large consumption of khalaal and rutab dates, these are usually put straight from the palm into rigid crates made of the dried midribs of palm fronds. The dates go to market without change of container.

The huge Basra crop, most of which is exported, is generally collected directly from the palms in wooden boxes; and it is in these containers that the dates are subsequently transported to the packing stations without further handling. Most of the dates never touch the ground, and those that do pick up remarkably little dirt, for the ground at harvest time is extremely hard and dry.

These boxes measure 24 x 28 x 49 cm, externally, and hold 1/30th of a cubic metre of dates, equal to 17.5 kg of loose dates and 32.5 kg of packed dates. This is the standard size for the export trade to the United States.

In Algeria and in Tunisia, much of the Daglat Nuur crop goes to the packing house on the bunch. For that part which is boxed, since the dates are not in the *tamar* stage, a shallower box is preferred.

FIELD GRADING

During harvesting there is usually some grading. *Hashaf* (shrivelled), *humeira* (which means the little red one), small, spherical dates, with small pit with a little dried flesh round it and papery red skin, and *khaiyis* (rotten) dates, are collected for cattle feed and thrown into a special basket. *Khalaal* are thrown into another basket, to be eaten fresh by the cultivator and his family, and the *rutab* into yet another. The latter will be spread in the sun to dry down into *tamar*.

Where the dates are delivered to the packing house still on the bunch, as is much of the Daglat Nuur crop in North Africa, there is not much grading in the garden.

In Basra, on the other hand, dates from the palms are usually carried in their boxes to a central collecting point in the garden known as a *jaukhaan*. If the crop is small and the packers are eager for dates, there is no further grading, but if the crop is large and packers are in no hurry to buy, then growers will often practice a second grading out of rejects in the *jaukhaan*.

In Baghdad, where most of the date crop is of the Zahdi variety, much is cut and carried on the bunch either as late *khalaal* or as *chasib*, which is a date that has dried straight from the *khalaal* stage without ever having been a proper *rutab*. In a Baghdad *jaukhaan*, therefore, there may be moderately thorough grading, with the *khalaal* left on the cut bunch to mature, the soft dates selected for stamping into baskets or pressing into skins, and the *chasib* picked out for sale loose. At the same time the refuse is thrown into a basket for the cattle.

On the coast of Libya, the dates are well graded in the gardens, first during the preliminary cutting in two; afterward, when the dates are turned over or collected for the night, bad dates are removed and given to the cattle. Moreover, during the growing season, windfalls are eaten from under the palms by the sheep or are collected by the children for the sheep's evening meal, so that there is not much admixture of rotten and wormy windfalls with the main harvest.

FIELD PITTING AND CAPPING

Pitting

Pit is the term commonly employed in the United States for a date seed, and stone in the United Kingdom. In the United States, therefore, packets of dates which have no seeds in them are marked "Pitted", but in the United Kingdom "Stoned".

Field pitting is generally the women's job. For the most part they pit in the gardens, but the work may be done on the house roofs where the dates are drying in the sun.

There are several ways in which the pit is extracted. In the United Arab Republic, when soft dates are being prepared for agwa, which is pressed into earthenware jars, the pit is squeezed out of the date between the finger and thumb and, at the same time, the skin is slipped off.

The teeth are frequently used for extracting the pit. In Libya, where the pits are removed when the khalaal are split in two before sun drying, the two operations are frequently performed at the same time by crushing the date between two stones.

In the United Arab Republic, in Makran and Iraq, a small knife is also used. In Iraq, the only pitting done in the gardens is that for the supply of packers who may not have enough labour in their packing stations to do all the pitting required before the departure of the next date steamer for the West. This pitting is of a high standard. The date is slit with the knife down one side, the pit is extracted, and the two cut edges of the date are put together again so that the cut is not visible. Care is taken to leave no white thaub (literally gown, meaning the paper-like endocarp which encloses the pit) protruding.

In Darab (Iran), field pitting is done with a blunt-ended needle set in a wooden handle. The pit is pushed through the date endwise.

Capping

The term "perianth removal" is clumsy, but "deperianthing" is worse. In the United States, the term "calyx removal" is found in the literature, which has the advantage of being one syllable shorter but the disadvantage of being incorrect. What is actually removed is the perianth; the calyx is only a part of that structure. Perhaps from cap-stem, which is a trade word for the dried date perianth, the words cap and capping may be made to serve by virtue of their brevity and tolerable correctness.

In some varieties of dates the cap usually adheres more firmly to the date than in the stalk, and in others more firmly to the stalk. Dates in the soft, rutab stage, particularly those of the Barhi variety of Basra, when pulled off the stalk, sometimes leave not only the cap behind but the pit as well.

Caps vary in the manner in which they are joined to the date. The Umm ed-Dihin date of Basra, for example, has a depressed centre to its base, so that the cap is sunk as it were in a pit. The Bureim and Istaamraan (Sayir) from the same place have flat bases from which the caps protrude, while the Umm ez-Zuweiwiida (the mother of the little bit extra) of Hartha has a little eminence in the middle of the base on which the cap sits. To prevent confusion in reading the above sentence, with its appearance of illogicality, it has to be remembered that the base of the date when it is young is at the bottom, but that as the date grows its weight pulls down the stalk on which it is borne, so that it ends its life hanging upside down, with the apex (when viewed from the ground) below and the base above.

When, in the gardens of northern Libya, the Khalaal date is sliced in two and the pit removed, the cap may fall away from the two halves and from the pit, or it may remain attached to one of the three. The slicer should remove caps adhering to the halves, but a large number nevertheless find their way to the packing house, where they have to be removed by hand on the grading belts. In the United Arab Republic the caps are removed in the gardens when the dates are pitted for the preparation of date paste. In Iraq, the caps are not removed before the dates are pressed into baskets; but where garden pitting is practised on behalf of the packer, the caps are, of course, carefully removed. When dates of the Zahdi variety of northern Iraq fall, or are pulled off the bunch, the caps nearly always remain attached to the stalks.

The caps are thrown with the pits during garden pitting and capping and fed to the cattle.

It has been previously noted that whether the date has or has not a cap, is of considerable difference to the packer. The reason why this is so is because the cap seals the base of the date; when the cap is removed a hole is left.

When the date has a cap it is unlikely to harbour a worm; when it has no cap it is often infested. When the date has a cap and the skin is unbroken, any dirt collected on the skin is readily brushed or washed off, but when dirt has entered through the cap hole, there is no practicable way of cleaning it.

FIELD RIPENING AND CURING

The terms ripening, maturation, curing, drying and dehydration require some explanation.

Each of the above, from ripening to dehydration, is considered to be a processing operation when it is carried out after harvest; when it occurs on the palm it is not considered to be so.

Ripening and maturation in ordinary speech mean much the same thing, but ripening gives more the idea of a continuing process, while maturation rather one of the completed process. Maturation, the grander word, tends to be more used in the packing house. For the field, ripening will do very well.

A ripe date is one in which the hard, brightly coloured _khalaal_ has been completely changed into the duller coloured, soft _rutab_. This is a squashy, perishable fruit.

It has already been mentioned earlier that some varieties of dates always, and others sometimes, skip the _rutab_ stage and go straight from the _khalaal_ to a tough, generally light-coloured, dry, long-keeping date known variously as _abul-khusheim_, _chasib_, _ariish_, _frezza_, etc. Here ripening and curing are telescoped into one process.

Curing implies a later process than ripening. The date ripens and then it cures. On the other hand, in the literature, it is often used indiscriminately with ripening, so that one finds articles on the home curing of dates when all that is meant is ripening. It is, however, a useful word to have for the further drying of the fully _rutab_ date into the prune-like, firmer _tamar_. The _tamar_ is not a perishable fruit; it is a dried fruit.

Drying and dehydration have the same purpose, namely, reducing the water content. In practice, however, drying of dates is the removal of moisture by natural evaporation, either in the sun or in the shade. In this operation there is little control or adjustment of the prevailing conditions. Dehydration, on the other hand, is the removal of moisture under controlled conditions of the temperature, the air speed, and the humidity of the air used in the dehydrator. It is to be noted that drying and ripening are not the same thing. The two processes, however, are intimately associated with each other, but they do not necessarily proceed at the same rate. They even proceed in inverse ratio during the earlier stages of ripening, that is, in the early kimri. The association, however, between drying and ripening in the later stages is much closer. A date can, nevertheless, be ripened without much loss of moisture, and it can be dried without being properly ripened.

Before considering ripening and curing in greater detail, six general statements are presented below. They indicate briefly regional climatic differences that affect these phenomena and the field in which artificial processing is useful.

1. A date properly ripened on the palm is of better quality than one picked before or after ripening.

2. A date picked before or after proper tree ripening may be processed and its quality improved, but its quality is never better than that of one properly tree ripened and is generally worse.

3. Most of the dates of the world are never submitted to any processing operation except pressing.

4. Where dates grow best, the climate is so hot that the ripening period is short and the dates ripen comparatively evenly.

5. In marginal lands, the climate is cooler and dates ripen more slowly. In such lands, two months may elapse between the ripening of the first and last dates on the bunch. In still more uncongenial climates, the dates may never ripen on the palm beyond the khalaal stage.

6. In a very hot but extremely humid climate, e.g. that of Bahrein or of Lingeh on the Persian Gulf, the dates form excellent rutab but may fall to the ground before reaching the tamar stage.

The ripening and curing of dates on the palm are the earlier and the later parts of a continuous process; but for the purpose of a review of the treatment the dates receive after picking, it is convenient to consider them in their different stages. Therefore, the treatment given kimri is first described and it is followed by that given khalaal, and so on.

RIPENING "KIMRI"

In Makran (Pakistan), a barren country where the ripening of the date crop is eagerly awaited, the poorer people often cannot wait for the khalaal and they eat the kimri. These are first put in a basket and beaten with a stick, through which treatment some of the astringent, tannin-loaded juice escapes. They are then placed in an earthenware jar, which is wrapped in a blanket and left for a night. In the morning, the bright green of the date skin has changed to a mud colour and most of the bitterness of the kimri has disappeared. This may be caused by the mechanical breaking down of the cells of the date with the consequent liberation of that enzyme responsible for the deposition of the tannin in an insoluble form.

Even so, these beaten kimri or sorapag (sour water), as they are called, are not thought much of by people who have other things to eat. In Basra also, in the pre-oil age, the kimri were treated in a similar way.

Kimri, when half grown or more, that fall from the palm to the ground, are sometimes bruised by the fall. The bruised part also turns mud coloured and is then sweetish to the taste.

In Makran, kimri, if they have nearly reached the khalaal stage, are sometimes exposed to the sun on the hot sand and will complete the change to inferior khalaal. They are then called pappuk rekiwar.

Neither sorapag nor pappuk rekiwar will keep for more than a day or two, because they are too moist and contain too little sugar.

RIPENING "KHALAAL"

Khalaal are ripened artificially for one or two reasons, either because the bunch has been cut when most of the dates were rutab or tamar and the others have to be treated in such a way that they also become rutab, or because the climate is such that the dates, if left hanging on the palm after the khalaal stage, would not continue to ripen properly or would be damaged by insects or rain.

As has been mentioned, the khalaal is a fresh fruit and large quantities are eaten in this stage, but the fruit will keep for only a few days. What cannot be consumed in that time has, therefore, to be dried or treated in one of the ways described below. It is useful to recall and summarize the information already given about the water content of the date in its different stages. Subject to reservations for particular varieties and for particular climates and seasons, the following are roughly the percentages of moisture found in the various stages:

	Moisture Content %
Earliest khalaal	85
Latest khalaal	50
Rutab tip	45
" 50 %	40
" 90 %	35
" 100 %	30
Tamar	20

Dates in all the above stages except the tamar are perishable fruit, but the less the moisture content the more days they will keep. Tamar dates are the only ones that will keep indefinitely.

Insofar as the khalaal is concerned, therefore, drying has to reduce its moisture from 50 to 20 percent, unless the dates are to be preserved, as they usually are, by being stamped into baskets so that air and water vapour are excluded, when a higher water content (25 percent) is in some circumstances permissible. Up to 30 percent may be allowed for the paste of pitted dates stamped into earthenware jars.

It is to be noted that a reduction of moisture from 50 to 20 percent requires that 75 percent of the original moisture be lost, as may be seen from the following example, in which the original weight of the dates if assumed to be 100 grams.

Weight of khalaal	100 g = 100 %
" " dry matter	50 g = 50 %
" " water	50 g = 50 %
" " tamar	62.5g = 100 %
" " dry matter	50 g = 80 %
" " water	12.5g = 20 %
50 g - 12.5 g = 37.5 g of 50 g of water	= 75 %

From the statement presented above, it can be observed that the earliest khalaal, that is, the date that has just changed its colour from green to yellow or to red, contains over 80 percent water. If picked from the palm at this stage, it cannot be turned into rutab. Rutab can be made only when the changes of colour from yellow or red to brown and of texture from firm to soft have begun at some point on the date, or when such changes are imminent. If these changes have begun, or are about to begin, they will continue, if the picked date is kept warm, until it is soft all over, that is to say, until it is fully rutab.

SIMPLE SPREADING IN THE SUN

This ripening process is practised in the gardens and on the house tops from Marrakech (Morocco) to Multan (West Pakistan). The khalaal are spread on mats one layer deep and exposed to the sun. At night and in rainy weather they are heaped up and covered, or brought into the house, hut or tent. Sometimes the rutab, as they form, are picked out and marketed, but usually the dates are left until they all become tamar or near tamar. Ripening and drying may last from three days to three weeks, depending upon circumstances, but a week is probably the most usual period.

In some places, such as the United Arab Republic, for example, drying grounds (called mansharaat, from nashara, meaning he spread out) with temporary matting walls are made in the gardens. On the south coast of Arabia, they are generally on the hard soil or rock outside the gardens and surrounded by permanent mud or rock walls 1.5 to 2.5 m high. They are known as duruua (singular dira). On the coast of Tripolitania (Libya), the dates are dried on mats on the ground at any convenient place in the gardens and are not protected by walls, either permanent or temporary. As much of the harvesting is done by the nomads who pitch their tents among the palms, they often prop up the eaves with sticks and spread their dates to dry in the slack.

In Mexico, the drying mats are usually on rough platforms about one metre above the ground; in Tibesti (Chad) a small platform, generally made out of the trunks of the duum palm and called angeli is used. The Teda (Libya) date owner usually takes his afternoon nap under it, thereby securing shade and guarding the dates at the same time; and in the mosquito season he will often sleep at night on top of the platform. Figure 36 shows drying stands and drying trays as used in Talesi Bagh, Muzaffargarh, West Pakistan.

BLACK SHEDS

A refinement of the sun-drying process has been introduced by some of the French Daglat Nuur growers of south Tunisia, who put the khalaal onto wire mesh trays, which they then stack in corrugated iron sheds painted a dull black on the outside. The temperature inside the sheds rises very high and the nightly labour of covering the dates or bringing them indoors is avoided.

SHRIVEL

Some of the khalaal spread out to dry will not ripen properly but shrivel. These are picked daily and fed to the cattle. Even khalaal on the palm will sometimes shrivel if exposed to a too dry and too hot wind. To prevent such shrivelling, the Yamani date farmers in the Toeg Hanbali near Djibouti, French Somaliland, wrap up their bunches of khalaal in rags and palm fiber as soon as the fierce, scorching winds of the khariif, the southwest monsoon, begin to blow. The farmers, in effect, thus retard rather than promote ripening.

In some places in Fars (Iran) the bunches of khalaal are also wrapped but with dried bushes. The protection thus provided is more to keep off the rain and only secondarily to keep out the hot winds.

SPLIT DATES

So far there have been considered khalaal which, when picked as such and spread out in the sun, will ripen into rutab and tamar without further treatment. These are the khalaal of lands where the sun is hot enough to bring about such ripening and where only the drop in temperature or heavy dews at night have to be guarded against; but there are other places where day temperatures are too low to allow the date to dry before fermentation sets in. Such places are the coast of Tripolitania (Libya), the island of Djerba (Tunisia), and the Punjab (Pakistan), where the khalaal will not dry quickly enough unless they are cut in half.

From a study of the literature, the permeability to water vapour of the unbroken skin of the khalaal between the stomata does not seem to have been determined, but it appears to have been assumed that it is virtually impermeable and that evaporation from the date takes place only through the stomata. As there seems no reason to doubt this assumption, it is clear that the cut surface of the flesh of the date must allow evaporation at a very much greater rate than through the skin of a whole date.

The labour of cutting each date longitudinally in two by hand, performed daily by the owner or tenant and his family, is enormous, but a moderately well-keeping date is produced when the halves, after drying for a week or so, are properly stamped into baskets. The great disadvantage of this slicing, however, apart from its employment of much labour is that, whereas sand blown onto the shiny surface of the whole date can be readily blown off again, when it blows onto the cut surface of the date flesh it sticks there, and the more the date dries, the more firmly does the sand adhere to it. The halves are dried on the ground and not on raised platforms, and since the ground is sandy and the drying grounds are not protected by walls, the amount of sand collected by the dates is enough to make them very gritty to eat.

Almost everywhere along the Tripolitanian coast the commonest date is the Bikraari, so that this is the date most usually found in the coastal markets in dried, slit halves, pressed into baskets.

On the island of Djerba (Tunisia), the Limsi dates which are not eaten fresh are also split and dried, though they are generally pressed into earthenware jars and not into baskets. The Limsi is the commonest variety there to be so treated.

At Muzzaffargarh in the Punjab (Pakistan), it is the Ghatti date which is the one commonly split before drying. The pit may or may not be removed. Dates dried in this way are known as <u>chirvi</u> or <u>shingistan</u> (compare the <u>tamar</u> of the Libyan coast).

CURING "RUTAB"

Ripening <u>khalaal</u> in the sun after removal from the palm is characteristic of the comparatively cool north coast of Africa, where the dates on the whole do not form good <u>rutab</u>, but the field ripening of <u>rutab</u> after removal from the palm is characteristic of the coasts of Arabia and of the Persian Gulf, where both heat and humidity are higher, where the dates of most varieties ripen on the palm into good <u>rutab</u>, but where many fall to the ground before the tamar stage is reached.

Thus, on the Baatina coast of Muscar, the drying ground (<u>mastah</u>), separated from the rest of the garden by a fence (<u>hadhiir</u>) made of dried palm fronds and palm fiber rope, is common. The drying grounds are beside the elaborate furnaces in which the dates are boiled to make the <u>khalaal matbuukh</u> which is exported to India and Pakistan.

Mabsali, Sillaani, Naghal and Qash Manuumah are the commoner of the varieties dried down from rutab to tamar. The commonest variety, Umm Silla (88 percent of the palms), does not need so much drying on the ground, perhaps because of its small size and high proportion of pit (19 percent compared with 13 percent in the Hallaawi of southern Iraq and may be stamped into baskets directly from the palm.

In Bahrein, the commonest date is the Marzubaan, which is generally harvested in a late rutab stage and dried for perhaps a week on mats. The drying is frequently not carefully done, the depth of the dates being too thick, possibly owing to shortage of labour. There is also loss from fermentation. The Khuneizi, the next most common date, withstands the humidity somewhat better but still needs some curing. There are, however, many varieties which are normally eaten as rutab, for example, the Khuwaaja, Maajiiya and Salmi, and no attempt is made to dry them down further into tamar, probably because they do not dry well. Hallau, on the other hand, though generally eaten as rutab, dries down to a good tamar.

In Qatif, the coast of Saudi Arabia opposite Bahrein, the Khuneizi is the commonest date and the Ruzeiz the next commonest. The usual practice is for the dates of both varieties to be harvested as rutab and cured for about a week on mats in the sun. The rutab of the former date need rather less curing than the latter, but the Ruzeiz cures into a better tamar.

It is evident that varieties differ in their capacity to produce good tamar, rutab, and even good khalaal, in unfavourable climates, so that by the selection of varieties, it may be possible to improve the general quality of the date yield in a given district. Another and quicker way to do this is by the introduction of improved drying methods, using either the heat of the sun, as in the black sheds of Tunisia, or oil-heated dehydrators, as in Libya.

Even in regions where dates ripen well into tamar on the palms, there is always a certain amount of sun drying of rutab to be done, because when the bunches are cut early in the season, not all the dates are hard enough to pack. Therefore, the softer ones, that is the rutab, have to be separated from the tamar and dried. The sooner this is done the better, because they then suffer less damage due to squashing. The least loss occurs when the separation is made at harvest time under the palm, when the tamar are put into one lot of containers, and the rutab, only one layer deep, into others. In the rush of harvest, however, complete separation is not always possible and the process may have to be repeated at the garden collecting point, and again in the packing house. However, by the time rutab are separated in the packing house, except those of the Daglat Nuur variety, which receive special treatment, as will be described later, they are too squashed over to make a first-class pack. Consequently, after drying on mats in the sun, they are generally packed in baskets for a low-price market.

In the Basra packing houses, the rutab that have to be removed during grading are usually in a late rutab stage and are ready to pack in baskets by the time they have lost about 4 percent of their weight, that is, about 17 percent of their water, as may be seen from the following calculations.

	Grams
Weight of rutab in sample	100
Less weight of pit at 13 %	13
Therefore, weight of flesh	87
Weight of water in flesh at 28 % before drying	24
Loss of weight on drying	4
Loss as percentage of flesh	5
Loss as percentage of water in flesh	17

The curing of rutab need not be carried so far if the product is to be made into a paste and packed so that air is well excluded. Thus the dastlagash, made from Haleni and Dandari dates in Kech and from Muzaati in Panjgur (both places in Makran, West Pakistan), consists of incompletely cured rutab, from which skin, pits and caps have been removed, kneaded into a paste before being tightly pressed into sheepskins or into small baskets made of the woven leaflets of the date palm. In the United Arab Republic the Haiyaani date is often similarly treated and packed in jars.

CURING "TAMAR"

In a dry, hot climate the dates dry into tamar on the palm. They are then no longer a fresh fruit but a dried one. When they are cut from the palm at this stage they may be packed at once without further drying, or stored in boxes until there is an opportunity for sale or packing. In the Sudan, however, where, if anywhere, one would have thought the dates as harvested would be dry enough, the ordinary practice is for tamar to be left lying in heaps on the sandy floor of the shuuna, the clearing in the garden where the dates are collected, for a month or two before they are sacked and marketed. This exposure to the sun dries the already dry dates further, but the object of the exposure is not, or not mainly, the drying but the killing of noxious insects, presumably Pyraloidea. Every week the heaps are turned by men with shovels, though the turning is less to ensure an even drying than to discourage attack by termites.

FIELD FUMIGATION

In the chief date-producing district of the world, that is, southern Iraq, the pests that attack dates on the palm, namely, the mite, the Lesser Date Moth, Batrachedra amydraula Meyr., and the Greater Date Moth, Arenipses sabella Hmps., have all left the palm before harvest. Therefore, if the bunches were cut, their dates put straight into boxes, and the boxes delivered to the packer before dusk, field fumigation for these dates would be unnecessary. There is, however, a circumstance which disturbs such a happy state of affairs. This is the practice of mixing together the dates which have been stripped off their stalks by hand into the boxes with those which fell to the ground during cutting and with the windfalls that were lying on the ground before cutting began.

The Fig Moth, Ephestia cautella Walk., does not lay its eggs on the dates on the palm, with the possible exception of those that have had their skin pierced by birds or wasps. The moth, however, oviposits on nearly every date lying on the ground without its cap (perianth), that is, on a date which has fallen from the palm, leaving its cap attached to the spikelet. The newly hatched larva of the Fig Moth cannot gnaw through the cuticle of the date and, therefore, makes its entry through the hole left by the absent cap. Early in the season the Grain Beetles, Oryzaephilus spp., are few; they attack the windfalls and stacked dates later, but every evening until the weather gets cold the Fig Moth and some of its cousins can be seen ovipositing on the harvested dates lying in heaps or in boxes.

On a large estate in Basra, Iraq, the practice was adopted of collecting all the windfalls from below the palms before the bunches were cut, and of keeping them apart from the insect-free dates shaken and pulled from the bunch. In this way the packing house was assured of receiving most of the crop from that estate insect-free. This practice, however, has not been copied, so that field fumigation for almost the whole crop is desirable.

In southern California, fumigation is generally practiced as soon as possible after harvesting, not because windfalls are mixed with good dates, but because Nitidulid Beetles infest the dates on the palm.

BUNCH TREATMENT

Lindgren and Vincent (1953) reported that the infestation by the Nitidulid Beetles, *Carpophilus* spp., that are found on the dates on the palm in California and that remain in the harvested dates, could be considerably reduced by dusting the bunches a few weeks before harvest with malathion and also with other insecticides; but as such dusting cannot properly be considered fumigation, it is no more than mentioned here.

METHOD

Tarpaulin

A tarpaulin, also called paulin, made of gastight material, can be used to cover a stack of date boxes or a heap of dates or, if the stack or heap is very big, several paulins can be used and joints made by tightly rolling the two edges together. The edges on the ground can be kept down with earth or sand, or with gunny bag sand snakes. The ground before stacking should be liberally sprinkled with an insecticide in powder form.

On top of a stack of boxed dates, it is advisable to have a gassing dome, that is, a space between the top of the stack and the paulin to allow space for gasification and to facilitate the diffusion of the fumigant. This is ordinarily made by four empty boxes arranged as a square with a hollow core. In the centre is put the tray to hold the fumigant if it is a liquid, or the pipe outlet if it is a gas.

A convenient size of stack is one of four tons of dates in 210 Basra-size date boxes, stacked seven high, because that height is the highest a man can conveniently reach when stacking. A paulin for a stack of this size, including allowances for the gassing dome and for a 50-centimetre flap all round on the ground, would measure 6.5 x 8.5 metres.

Paulins of canvas are being rapidly replaced by those of lighter, plastic material.

A nylon paulin coated with neoprene and of the above dimensions is expensive. A much cheaper device, and one that is quicker to put over and to remove from a stack, is a box consisting of a light, wooden frame, covered with strong, bituminized paper. The bottom of the box is open. The paper projects round the bottom edge, so that when the box is in position over the dates, the flap can be weighed down on the ground with earth to reduce the escape of fumigant. One of the longitudinal members of the box frame on each side projects at each end to form handles. A box to cover four tons of dates measures 3.5 x 2 x 2 m and does not weigh more than 80 kg. Four men, one at each handle, easily raise the box shoulder-high and then lower it over the stack of cases. It has, however, to be recognized that unless the material is gastight, it will be difficult to maintain the necessary concentration of the fumigant. Impregnated canvas is a better material than bituminized paper but is more expensive.

Mud Chambers

Winter (1959) advocates the use of mud-built fumigation chambers in the field. They have much to commend them in regions where clay is at hand and where house construction in mud is practised. They can be made reasonably airtight if well plastered with mud from time to time, inside and out, but they suffer from the disadvantage of difficulties in sealing the door and of danger if, after the door is opened, enough time is not allowed for the dispersal of the fumigant before the entry of the labourers to remove the dates.

Fumigant

Field fumigation differs from packing house fumigation in being practised by farm labourers and other persons without the supervision of skilled technicians conversant with the dangers in the use of powerful fumigants. Therefore, for field use only, a fumigant that is comparatively harmless to human beings is recommended, even though it may be less efficient than others. Such a fumigant is a mixture of three parts of ethylene dichloride and one part of carbon tetrachloride, called ECM or chlorasol. It has been used successfully for many years by a large date company in Basra, Iraq, by unskilled people, and without casualties.

Kapsiotis (1956 and 1958) states that this fumigant does not destroy insect eggs, so that dates may need more than one fumigation before packing in order to kill the young larvae that hatch from the eggs unkilled by the first fumigation.

Both Kapsiotis and Winter (1958 and 1959), therefore, recommend methyl bromide for field fumigation. They state that the dispersion and dilution of this gas in the open air is so rapid, that there is no danger of poisoning if precautions are taken. Kapsiotis adds that chlorasol is five times as expensive as methyl bromide. Winter, however, points out that it is necessary to recognize that the use of methyl bromide and many other insecticides under field conditions always presents danger, and that they should, therefore, be handled with the utmost care and only by trained men under responsible supervision.

FIELD PACKING

Most of the world's dates are packed in the field and only a small part in packing houses. It is likely that field packing will continue to be important in the date industry because, although packing house packing is cleaner and in several ways better, it is more expensive. Field packing is largely for the cultivator and his family, but dates packed in the field also enter widely into commerce, such as that practised in Basra for the trade in low-quality dates.

The containers used in field packing, with two exceptions, are made by the cultivator and his family, or by his neighbours in the village, from local materials. The exceptions are the modern 18-litre kerosene or petrol tin and the 200-litre oil drum. The earthenware jar is also in some places an exception because it must be brought from a distance if there is no suitable clay.

Dry dates are packed in sacks, but the cultivator will press soft ones into a compact mass in a suitable basket or other container; such a compact mass keeps reasonably well. The materials available to the cultivator for containers suitable for these purposes are the leaflets of palms, the skins of sheep and goats, earthenware jars and, by stretching the meaning of container somewhat, strings may be included.

When a container is small, for example, a goatskin, the dates are rammed down tightly with the clenched fist, but for bigger containers, in the absence of mechanical presses, the feet are used.

It is, nevertheless, hardly to be doubted that the packing house with its cleaner methods, despite their cost, will increase its production at the expense of the field pack.

BASKETS

Basket making is one of the oldest crafts and the leaflets of the date palm lend themselves readily to it, so that it may be supposed that they have been used for date baskets from remote times. What appears to be the earliest mention in English of baskets of dates, as opposed to mention of dates only, occurred in March 1631. It is in a letter to the East India Company from Richard Barry, who mentions loading "frayles" of dates in the Persian Gulf. It is, however, to be expected that a study of early Arabic literature, or of the Sumerian clay tablets, would bring to light very much earlier reference. The "frayle" is the farsala, the large date basket of the Gulf of Oman and the Hadhramaut coast.

Baskets to be used as date containers are most commonly made from the leaflets of the date palm, but on the coast of Libya, where esparto grass (*Stipa tenacissima* L.) is common, this is used also. It makes a stronger but more costly basket. In the Sudan, the leaflets of the duum palm (*Hyphaene thebaica* Mart.) are also used but those of the date palm are preferred. In Makran (Iran and Pakistan), baskets made from the leaflets of the date palm are used but those made from the Daaz or Mazari, the dwarf palm (*Nannorrhops ritchieana* H. Wendl.), where it is available, are preferred.

The younger the leaflet the greater is its pliability; but it is also narrower and shorter. Thus, the strip of matting into which the young leaflet is made to narrow, and the time required to make a given area of strip is longer than it is with older leaflets. The leaflets are split in two longitudinally and plaited together to form a long strip eight cm or so wide. The strip is then coiled on itself in a spiral and the edges sewn together with string made from leaflets rubbed between the hand, or cut off the well-beaten and then split main stem of the date fruit bunch. When the strip has been wound round in a spiral flat on the ground to make the bottom of the basket, the winding continues but the strip is now set on edge so that the side of the basket is made. The top of the basket is usually narrower than the bottom.

SKINS

Skins which are no longer serviceable as floats, receptables for water, milk, sour milk, honey, or as milk churns, are used as containers for soft dates. In the climate of Baghdad (Iraq), where they used to be employed in large numbers, especially for the Khastaawi variety, dates keep in them excellently for a year or more; but Munier (1955), writing of the dates in skins in Mauritania states:

> "The conservation of dates in skins is only possible in very dry regions, thus barring exports to territories with a humid climate, such as South Guinea, Ivory Coast...".

Skins, however, still maintain much of their importance in those date countries where grazing occupies a large part of the population, namely, Mauritania, Algeria, Chad, Sudan, Saudi Arabia and Makran (Pakistan).

JARS

The origin of the potter's art, like that of basket making and of curing skins, must be extremely ancient but, here again, the earliest noted reference in English to dates being preserved in jars is of comparatively recent date. It is that of the anonymous narrator of Vasdo da Gama's first voyage (Ley, 1497). He states that some of the crew were entertained by the lord of a part of the Island of S. Jorge, off Mozambique, and given on their departure a jar of bruised dates made into a preserve with cloves and cumin. This was in March 1498.

Some of the unglazed and roughly glazed jars (Khaabiya, plural khawaabi) used for storing dates in the Fezzan, Libya, may be as much as 1½ m high and 1 m in diameter at the widest part.

FIELD STORAGE

As has already been mentioned, the greater part of the world's dates are packed in the field. Similarly, the greater part is stored in the field, the house, the tent, the mat hut or the local merchant's warehouse. The modern refrigerated or fumigated insect-proof store houses only the smaller part of the crop. Consequently, a short survey of the methods of field storage is appropriate. The practice of field storage may be expected to continue, so that it is important to know what methods are commendable.

In the previous section, "Field Packing", there have been mentioned baskets, skins, hars, drums and tins. All of these, with the exception of the bigger jars and drums, are portable and between them carry most of the dates in the world's trade, and all of them, of course, serve to store the dates as well as to transport them. The nomad stores his year's supply of dates in baskets or skins, the merchant in baskets stacked in his warehouses. The settled cultivator, however, for his home supply, does not need the container to be portable. He employs for temporary storage the not very satisfactory large, circular bins of matting, as in the United Arab Republic and Mexico, the <u>bunques</u> of Lower California (Nixon, 1953) or, for longer storage, the excellent large earthenware jar and rather less excellent mud or plaster bin.

FIELD TRANSPORT

Under this heading are considered the transport of dates from the palm to the collecting station in the garden or house, and from the garden or house to market or packing house.

A date garden usually has a clearing conveniently near a road, truck or waterway, and generally shaded by a mango, East Indian walnut, jujube, tamarind, mulberry, bead or other good shade tree. Big gardens may have more than one such clearing. In this shady, open space called <u>jaukhaan</u> in Iraq, the dates are heaped as harvested. To it the dates are conveyed in small baskets called <u>zanbiil</u>, carried on the heads of the women, girls and small boys, and in bigger, soft baskets called <u>jilaal</u> (singular, <u>jilla</u>), or in rigid baskets made of the midribs of date fronds, called <u>rukuuk</u> (singular, <u>rukk</u>), borne on the backs of the men and youths. The porters help support the weight of the last by means of a thong plaited matting over the forehead. In places where garden boxes, that is, lug boxes are used, or where export boxes are used as lug boxes, they are carried on the head.

In places where the cultivator's house is not in the garden and where the grading and packing take place in the house as, for example, in the Hadhramaut, in Mexico, and also in other places, the dates are carried from the gardens to the cultivators' homes by camel or donkey. The dates may still be on the bunches or they may be loose. They are tipped onto a sacking mat, the corners of which are then tied together, and two such mats, one on either side, are then loaded onto the pack animal.

Improvements on these more primitive methods of garden transport are the truck of California, the punt on the irrigation channels of Basra, and the Decauville railway in one garden in Basra and in a few gardens in the Daglat Nuur districts of North Africa. All these conveyances take the dates in wooden boxes. In the Daglat Nuur districts of Africa and of the United States, a shallower box is used than elsewhere, because the dates of this variety and of the Aliig are normally harvested at an earlier stage than those of other varieties, and are, therefore, softer and do not withstand crushing.

TRANSPORT TO MARKET OR PACKING HOUSE

The wooden box is the best container for dates being carried from the garden to the packing house. The lug or garden box, of stout construction, without lid, is normal in the fruit trade and is used for dates in the United States. In Libya also, in order to have the coastal, sun-dried Bikraari pitted halves brought to the government packing house in as clean and unsquashed a condition as possible, similar lug boxes for transport from the market were used, but the projected extension of the system to the gardens was not effected, because the packing house switched its demands from coastal dates to the desert variety.

Lug boxes, which are also the export cases, are used in Algeria and in Tunisia for the transport of the Daglat Nuur and Aliig from the gardens to the local packing houses and also to the packers in Marseilles and in Italy.

Similarly in Basra, the export cases are used also as garden boxes. Kapsiotis (1958) considers it undesirable that the one box should be used for both purposes, and for the modern export of dates in small, cellophane-wrapped packs, this is so. For the still considerable trade in tightly pressed dates in bulk cases, it would not seem to matter much if the export case has also collected dates from the gardens because, during pressing and during the voyage to Europe, America, Oceania or South Africa, the date syrup from the packed dates in any event exudes and stains the box. Moreover, using the same box for the two purposes cuts the overhead on each operation by half.

Fibreboard cases with a capacity of 15 kg of dry dates each are now being used for the government date-packing house in Libya because, as mentioned above, the packers now buy Fezzani dates instead of the coastal Bikraari. The reason for the change is that, on the one hand, it has proved impossible in the packing house to remove all the sand from the sun-dried, pitted halves by washing, while on the other hand, school children in the towns, for whose school breakfasts the dates are destined, have become fastidious in respect of them. In the Fezzan, the dates dry completely on the palm; they are not pitted and, if care is taken to grade out those with cracked skins and those without caps (**perianths**), a sandless date is obtained with very little cleaning. The date groves, however, are 1 000 km distant over bad roads, and the trucks which go down are loaded with goods that pay a high freight. On the other hand, the trucks ordinarily return empty, so that they will carry dates at a low freight. A container which weighs little and which when empty occupies little space, is therefore needed for sending down from Tripoli to the Fezzan. The flat-folding, fibreboard case has, therefore, been adopted; but it has the disadvantage of becoming easily damaged when not tightly filled and well sealed. Nevertheless, 1 000 t of dates were brought up from the Fezzan to Tripoli during the 1959 season in cases of this kind. Their use certainly keeps the dates in much better condition than that of the dates that used to be merely loaded into lorries without any containers at all.

The punts of Basra have already been mentioned, but on the Nile also, the local craft are used to carry the dry dates of Upper Egypt downstream to market. Most are heaped loose in the boar like so much gravel, but dates of the Gundeila variety are almost always sacked (Brown, 1924). The Daglat Nuur and Aliig varieties from the Sahara and Tunisia, which go mostly to France and Italy for packing, are carried, of course, by ship across the Mediterranean.

The train carries the Euphrates Zahdi to the Baghdad packing house, and also the Daglat Nuur and the Aliig to the North African ports; but everywhere the truck is supplanting the train for this class of business. Trucks are even generally preferred for the long haul from the port of Naples to the packers in Genoa, Savona and Albenga. From the garden to the local packing houses in the Sahara and Tunisia, the donkey cart, nowadays provided with rubber tires, is much employed.

It is not necessary here to describe the many kinds of baskets, panniers and sacks used for taking dates to market, but one ingenious device used in Musca may

Telescoping booms with metal baskets. California, USA

be mentioned. This is the Mishaan, a circular cylinder about 30 cm long and approximately 15 cm in diameter, made of pieces of palm frond midribs, tied together with palm fiber twine. One end of the cylinder is plugged with palm fiber, the cylinder is filled with dates in the kahlaal (yellow) or rutab (soft) stage, and the open end of the cylinder is then plugged with more fiber (liif). About 20 such cylinders, each holding about 3 kg of dates, are then packed in a long, wide, squat basket of woven palm leaflets (khuus) known as a marhala. Two of these is a donkey load. The dates arrive in the market at Matra (Muscat) in these containers in excellent condition after their three or four days' journey down from the mountains.

MECHANICAL HARVESTING IN CALIFORNIA

The desert dweller who owns a few oasis palms for his family's use may still use the age-old methods of climbing his tall palms with a rope looped around his back and the trunk of the palm to pollinate the bunches in the spring and climb again in the fall to harvest the crop by either cutting and dropping the bunches to the ground or by shaking the dates onto a cloth suspended under the bunch and lowering them in a basket by rope. His operation is complete after pressing the dates into hand woven baskets or earthen jars and storing for future use throughout the year.

For the mechanized scientific date grower who must get maximum production from his palms to enable him to stay in business such simple methods can no longer be employed, especially where production is high with ten to 20 thousand pounds per acre and where high labour costs and tall palms have forced him to mechanize his operations in order to cut his costs per acre or per pound of dates delivered to the packing house. It can be seen that one cent per lb reduction in harvesting costs on one million lb is $10 000 saved or earned, and in the case of the mechanical harvester costs have been reduced by over one cent per lb.

A date grower in the Coachella Valley, Southern California, USA, must plant his palms in straight rows at a set distance of 24 - 30 ft to allow his equipment to operate efficiently. His palms should be of the same variety in each plot. The land must be level to allow beneficial use and equal distribution of irrigation water. If his land has a high water table he will be required to install tile drains to a depth of 7 ft to remove salt from the root zone. The soil is analyzed in a laboratory to determine deficiencies which can be corrected with the addition

Baskets emptying into the hopper for removal of dates by the shaker. California.

of fertilizers and minerals as recommended by the soil scientist. Insect pests are controlled with insecticides applied by machine. As the palms grow to greater heights, machines are required to elevate the workers to enable equipment with trained operators for use by its growers. This equipment consists of mechanical harvesters with shakers, tractors with equipment for discing, bordering and furrowing, trucks and tractors mounted with dusters for applying insecticides, oil spray rigs for weed control, fertilizer injection equipment and trucks for hauling supplies and equipment.

Cal Date Company's double unit mechanical harvesters can harvest 40 000 lb of dates per day or approximately 200 palms. These harvesters consist of two fifty-foot telescoping booms with metal baskets attached to the end of each boom. Each of these mechanical harvesters is operated by a crew of five men, two drivers-cum-boom operators, two men in baskets and one man on the shaker. The booms are mounted on an international four-wheel drive four ton truck with an attached shaker which removes the dates from the bunch. Boom operators stationed on the bruck bed control the direction of the booms. A man is stationed in each basket and elevated to a position under the bunches at the base of the palm fronds. These men severe each bunch of mature dates from the fruit stalk with a sharp, heavyyknife and place the entire bunch in the basket. When all the bunches have been removed from the palm the boom is lowered to a position over a hopper near the shaker. A trap door in the bottom of the basket is released allowing the bunches of dates to empty into the hopper from where they are conveyed to the shaker. Another method using less costly equipment is to use an aluminium extension ladder instead of the hydraulic boom. Men climb the ladder, cut the bunches and lower them to the mechanical shaker by rope.

Several methods of removing the dates from the bunch by shaking or vibrating have been developed. The system used by Cal-Date mechanical harvesters utilizes the vertical shaker which incorporates a crankshaft, cylinder and piston having a three inch stroke. The crankshaft is driven by a hydraulic moter at 800 rpm. Machines are used to economically carry out the cultural practices such as dethorning, pollinating, dusting with insecticides, pruning, tieing down and thinning the bunches, bagging for rain and bird protection and, especially, harvesting. They are also used to chop the pruned leaves and bunches and return them to the soil as humus.

The date industry in California is constantly experimenting with the assistance and cooperation of the Date Administrative Committee, the University of California, large companies and the ingenuity of individual date growers and packers. Costs of equipment vary greatly from the custom built hydraulic extension towers owned

Harvesting with ladder and rope, and use of mechanical shaker. California.

by the large operators to equipment constructed in the workshops of the growers utilizing engines and other parts available at low cost.

Because of the scarcity of local labour or the reluctance of labourers to climb palms, California growers have been spurred into action to develop new machines for harvesting and other tree operations. Several mobile harvesting towers have been in operation for many years, but it was evident that more efficient methods, including mechanical harvesting would have to be developed. Evaluation studies conducted by Roy Nixon of the U.S. Date Experiment Station indicated that there was very little difference in quality when dates were harvested in one pick rather than in several pickings over the season. These findings made feasible the use of a mechanical harvester which would handle the bunches of tree-ripened dates in one efficient operation. The dates could then be allowed to mature on the bunch, using insecticides to control infestation.

Cal-Date Company of Indio, California, which handles a large percentage of Coachella Valley's date production, maintains a fleet of specially designed mechanical shakers.

A special hydraulic clamp attached to the piston grasps the fruit stalk firmly. The operator then activates the hydraulic motor which allows the piston to quickly reach the optimum speed of 800 strokes per minute. Within three and one half seconds all the dates have been removed from the bunch and have fallen into a bulk bin under the shaker. These bulk bins are conveyed into position under the shaker on rollers. Each bin is 44 x 44 x 20 in deep and contains 800 to 1 000 lb of dates. When full, the bins are fork-lifted onto a truck and transported to the packing house for fumigation and further processing.

It is important that dates harvested in this manner and stored in bulk bins of 800 lb capacity must be mature and well cured to a fairly low moisture content. It has been found that dates harvested in too soft condition tend to mash or become sticky with syrup. However, with a majority of dry dates and a few soft ones the whole bin increases in quality as moisture from soft dates is absorbed by the dry.

Note: Dowson points out that for machines like those described above to be profitable, at least ten requirements must be satisfied as follows:

1. the date palms must be widely spaced;
2. the palms must be planted in straight lines;
3. the palms must be adult;
4. the palms must be of about the same age;
5. the palms should preferably be all of the same variety;
6. there must be no interplanting;
7. there must be no deep, open drains or water channels to be crossed;
8. the yield should be heavy, if the machine is to pay;
9. the dates must be dry or nearly so;
10. labour must be expensive and machinery relatively cheap.

Most plantations in the Old World do not fulfill all of these requirements yet because of the high cost of labour in many of the date-producing countries a change of harvesting methods is urgently required. It is apparent that further study of the problem is necessary.

Reference may also be made to the papers by Perkins and Brown (352, 353) and by Wright (466) for progress in the mechanization of date harvesting in California.

CHAPTER XI

PESTS AND DISEASES OF THE DATE PALM 1/

This paper reviews the pests and diseases of the date palm on a worldwide basis and includes all literature available through May 1968.

The first two sections consider mainly the insect and mite pests that have been of concern to date growers during the past 20 years. Distribution, economic importance, injury, description, life history and control are discussed. Fumigation and packinghouse pest control procedures are described in the third section. Excellent reviews of date palm pests by Lepesme (250) in 1947 and Stickney et al (413) in 1950 provide a basis for the discussions. New information published since then is emphasized, but some reports prior to 1947 are cited when more details are desirable.

Although methods and materials for the control of various insects and mite species change rapidly, current information for specific pests is included. The common names of insecticides and acaricides are those approved by the Committee on Insecticide Terminology of the Entomological Society of America. No proprietary names are used for insecticides.

The relatively limited information on nematodes is reviewed in the fourth section. Diseases caused by fungi and unindentified factors are considered in the final section. Neither bacterial nor viral diseases have been definitely associated with date palms, and well-defined mineral deficiency disorders are unknown.

1/ This Chapter which appeared originally in the FAO Plant Protection Bulletin, Vol. 16, Nos. 5 and 6, 1968, has been contributed by H.S. Elmer, Department of Entomology, University of California, Riverside; J.B. Carpenter, U.S. Date and Citrus Station, Indio, California; and L.J. Klotz, Department of Plant Pathology, University of California, Riverside, U.S.A. Where additional notes on minor mite and insect pests, birds and mammals, from V.H.W. Dowson, have been inserted, this is indicated in the text.

Date palms demand large quantities of water for growth and fruiting in the hot deserts where they thrive, yet atmospheric moisture is their commonest enemy. Most of the important pathogens of palms are fungi and, except for the destructive soil-borne species of Fusarium that causes bayoud, are dependent upon high humidity and rain for their development. Thus, except for bayoud, the fungal diseases of date palms are of only sporadic importance in the more favoured date-growing districts, but are increasingly common and damaging in date culture, as conditions become suboptimal to marginal because of poorly distributed or excessive rainfall and high humidity.

The economic importance of date diseases is emphasized and estimates are reported wherever available on the losses caused by diseases, but few statistical data exist. The distribution, causal agent, symptoms, description of the causal organism, and control measures for the diseases are given where available or useful.

More than 30 years (see foot note) have elapsed since Chabrolin(84A) and Fawcett and Klotz (166) published general reviews on date diseases. Viennot-Bourgin (444) reviewed the fungal diseases of dates. Recently, Nixon (326, 333) summarized the information on diseases of dates in the United States and Calcat (72) discussed date palm diseases in the Sahara and North Africa. In addition, valuable discussions of environmental factors influencing date culture were contributed by Caty (80) and Péreau-Leroy (351). The review of date diseases here is an amplification of a 1966 report (79) originally prepared at the request of FAO.

1. MITES

Banks Grass Mite

Distribution and Economic Importance

Banks grass mite, Oligonychus pratensis, commonly called date mite in Californi is a major pest of date palms in the United States (146). This mite can cause extensive damage by ruining entire bunches of fruit, particularly those of the Deglet Noor variety which constitutes over 85 percent of the California date plantings.

Injury

O. pratensis rasps the leaf surface and causes affected areas to become greyish white or pale yellow. Infested green fruits may be scarred so badly that they turn brown. The skin on infested fruits becomes hard, and then cracks and shrivels, reducing the grade of the fruit. A heavy deposit of fine webbing is spun over much of the feeding area. This webbing collects dust, and even a moderate infestation in tall palms is easily observed from the ground.

Description and Life History

The biology of this mite and its host range were determined by Stickney et al (413). Pritchard and Baker (363) listed the host plants, which include many grasses in the United States.

The body of the adult female is ovate and about 0.29 mm long. The colour varies from yellowish white to pale orange, with more deeply shaded irregular spots along the side of the body. Adult males are bluntly subtriangular and smaller than the female, about 0.25 mm long, but of the same colour. The egg is pearly white, spherical, and approximately 0.13 mm in diameter. Immature mites resemble adult females except that the larvae have only six legs; after moulting they become eight-legged nymphs. The first infestations appear in early June when the fruits are small; peak populations occur on fruits in July and August. These mites usually do not remain on ripe fruit but maintain themselves during the winter months on late-ripening or off-season green fruit and also on the foliage of the palm or on grasses, particularly Bermuda grass, *Cynodon dactylon* (L.) Pers., growing in or near the date gardens. Although the number of mites decreases and their activity lessens during the winter months, all stages of development are present throughout the year.

The mites are incapable of flight and do not spin silk strands for the express purpose of floating on the wind as some mites do. Their main means of dispersal, besides walking, is probably transportation by wind. Strong, gusty winds are frequent in the date-growing regions of the United States and probably contribute to the rapid reinfestation that so often occurs in date gardens following successful initial control measures in early summer (146).

Control

Sulphur dust is used almost exclusively to reduce injurious populations of the Banks grass mite. Motor-driven dusters with tube extensions capable of reaching high bunches are commonly employed to deposit a light coating of sulphur over the entire surface of each fruit. From 25 to 50 lb of sulphur are used per acre (45 to 50 palms). Some growers prefer one sulphur dusting applied by aircraft before the date bunches are tied down, followed by a later application from the ground when mites appear on the bunches. The best mite control is obtained by blowing the sulphur directly upward into the bunches (445).

Poor coverage and subsequent poor control result when either the equipment is incapable of blowing the dust to high date bunches or when dust is applied during excessively windy periods. Since sulphur dust is irritating to the eyes of persons applying it or working in dusted trees, experiments were conducted to determine whether sulphur and other acaricides might be applied as sprays (143). Wettable sulphur sprays were less irritating to the eyes than dust and more easily applied to high date bunches. The bunches were thoroughly drenched with spray at the rate of 200 to 300 gallons per acre applied with a conventional high-pressure citrus sprayer and a shade-tree gun operated by personnel on the ground. Population levels were reduced sufficiently to prevent injury when six to 10 lb of wettable sulphur per 100 gallons of water were used. Spreaders and stickers, added to dust and sprays, hold some promise for more lasting control. Wettable sulphur sprays were used commercially in a few date gardens during 1962 (145).

Other acaricides tested as sprays or dusts were more expensive than sulphur and no more effective; none is recommended in the United States, as they have not been registered for use on date palms.

Old World Date Mite

Distribution and Economic Importance

The Old World date mite, *Oligonychus afrasiaticus*, has long been known as a serious pest of date palms in Iraq and North Africa. Its continued importance has been reported in North Africa (72), Libya(267, 268)and Iraq(243, 161). P. Munier 1/, who during the past several years has published numerous surveys on date culture in countries bordering Saharan North Africa, noted the presence of the Old World Date mite wherever dates are grown. In Libya (268), this mite is a serious pest of date palms in inland oases but not in coastal areas.

Injury

O. afrasiaticus is similar to the United States species, *O. pratensis*, and may cause the same type and degree of injury. In North Africa the Old World date mite is reported to be especially abundant on unthrifty and neglected palms. Nixon (157) summarized observations on the tolerance of several date varieties to infestation.

1/ Reports on date culture in Chad, Mali, Mauritania, Niger and the Sudan, published by the "Institut des fruits et agrumes coloniaux", Paris, France.

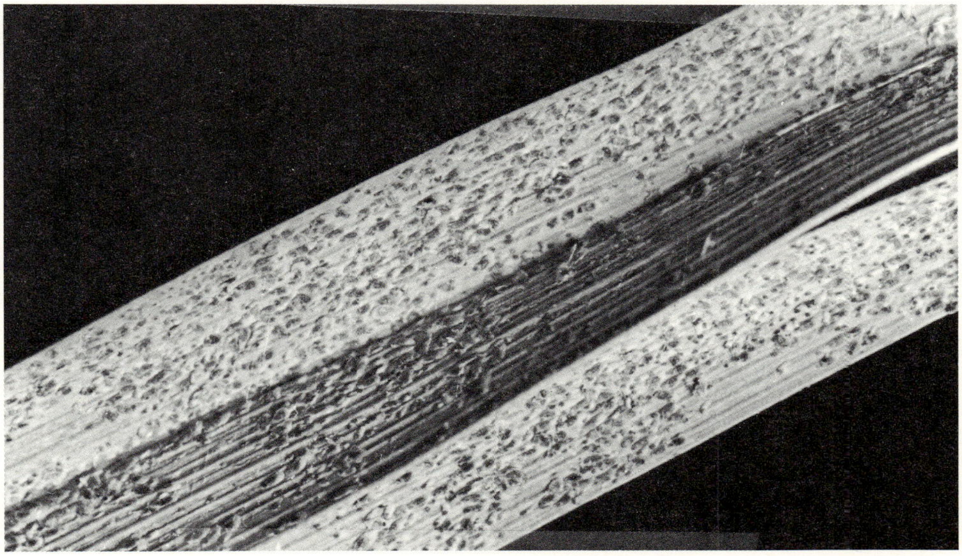

Parlatoria blanchardi

Nomenclature

Oligonychus afrasiaticus (McGregor), originally described in 1939 in the genus Paratetranychus, was referred to Oligonychus by Pritchard and Baker (363). Previously, this mite had been confused with O. pratensis (Banks) which has been collected only in the United States (146). Some literature still refers to the Old World date mite as P. simplex (Banks), a name listed by McGregor(275).

Control

Calcat (72) reported good control with a dust containing 25 percent sulphur and 75 percent sifted (hydrated ?) lime. In Iraq (161), Laville (243) recommended dusting with sulphur at the rate of 100 g per palm.

Mites in Stored Dates

Many species of mites can be serious pests if once established in stored dates. Infestations are rare, however, and can be controlled by fumigation (413).

2. INSECTS

Parlatoria Date Scale

Distribution and Economic Importance

The parlatoria date scale, Parlatoria blanchardi, is widely distributed throughout most of the date-growing regions of the world (91) except the United States. This scale insect is probably indigenous to Iraq (72) and spreads to new areas as the palms are transplanted by offshoots. If allowed to go unchecked, parlatoria date scale may become the most serious date pest in many areas. Losses of 70 to 80 percent of the fruit by direct attack of this scale insect have been reported by Smirnoff (411), who published the most recent and comprehensive review of this pest, including his own extensive investigations. Other reports include those from Libya(267,268), Algeria (424), Israel (90), Tunisia (343), Morocco (351), North Africa (72), and other countries (see footnote under P. Munier).

Injury

Under favourable conditions, parlatoria date scale can spread over the entire surface of the foliage and fruit of the date palm, covering them with live and dead insects. The scale feeds on the white succulent tissues at the base of the leaf stalk - one of the most protected and inaccessible places on the palm. As the populations increase they move from these areas on to the pinnae. Heavy infestations on the pinnae cause them to wither and die. In commercial plantings date palms are seldom killed by parlatoria date scale, but feeding by these insects definitely affects the vigour of the palms and decreases yields. Infestations on the fruits reduce their commercial value and may render them unfit for human consumption.

Description and Life History

Descriptions and life histories of the various stages of the parlatoria date scale were given by Stickney et al (413) and Smirnoff (411). The adults average about 0.7 mm long with the female being broadly oval and somewhat flattened in all stages. The male, after five nymphal stages, emerges as an adult with well-developed wings. The female adult remains stationary for life and produces numerous eggs. After the eggs hatch the nymphs crawl out from under the female and move about, feeding at various positions until the third moulting. These nymphs are pink to red in colour; female adults are grey. The egg period varies with temperature, ranging from two to 11 days. The generations are not clearly defined; there is an almost continuous overlapping of egg, nymph and adult stages throughout the year.

Control

Each date-growing area has specific problems of control, but a concentrated effort generally will reduce population levels so that injury is negligible. Martin(267,268) found that both diazinon and parathion gave satisfactory control in Libya. Diazinon would be preferred because parathion is too highly toxic for use by unskilled labour in most areas. In Israel, Kehat and Swirski (228) tested several materials for the control of this scale and found dimethoate the most effective. However, they also showed that insecticides applied to the date palm or to adjacent crops interfered with biological control of parlatoria date scale.

This pest was introduced into the United States in 1890 with the first successful importation of date palm seedlings from Algeria and the Arab Republic of Egypt. Efforts were made to eliminate the scale at that time, but it proved to be very difficult to eradicate with the techniques then available. An elaborate system of inspection, quarantine and control, by stripping the leaves and burning, was initiated around 1914. The last parlatoria date scale was observed in 1934. Apparently, this scale has been successfully eradicated from the United States (56).

Smirnoff (411) investigated this scale insect intensively, especially its biological control. The species of predators include: Sarcoptidae (Acarina), 1: Coccinellidae, 26; Nitidulidae, 5; Chrysopidae, 1; Mycetaeidae, 1; and Anthocoridae, 1. In a typical situation, a six to eight-year-old palm might be host to 30 to 50 million live parlatoria date scales and a predator population of 8 000 Nitidulidae and 3 000 to 4 000 Coccinellidae (*Scymnus* sp. and *Pharoscymnus* sp.). Another 100 million date scales might be present in the form of crusts on leaves and other parts of the palm.

The hand-shaker bellows used with a nicotine dust against the Date Bug. Basra, Iraq.

Predators are strongly influenced by climatic conditions and populations vary widely in component species and total numbers. In a single large oasis the distribution of parasites will be governed by a wide range of micro-climates. Substantial diminution of Parlatoria date scale populations has followed some parasite introductions. Biological control is especially valuable for those areas where consistent chemical control is impractical. In newly infested areas the scale may build up rapidly and become very damaging, but when predators are introduced under favourable conditions the scale populations are reduced to tolerable levels. Recent studies on biological and chemical control of this scale insect were reported in 1966 by several investigators (161).

/ Commenting on the above, Dowson notes that the Parlatoria scale, which was eradicated in the United States, only sometimes does much damage in the Old World, though Balachowsky (30,31) found it so serious in the western oases of Algeria that he endeavoured to import predators there from the east. Heavy infestation on the outer fronds reduces the length of their life and, therefore, reduces the amount of the crop.

It is possible that the general lightness of the damage to the date palms of the Old World by Parlatoria is due to the high degree of parasitism to which it is subjected. Both Mr. Laing of the Commonwealth Bureau of Entomology, London, and Mr. Magness of Beltsville, remarked on the large proportion of parasitized specimen in the collections received from Somalia; and Balachowsky reports that Cybocephalus palmarum, Peyerh., C. flaviceps, Rt.., and Pharoscymnus anchorago, Fairm. are predators in North Africa. In Somalia and Iran, ladybirds eat this scale with avidity. Tourneur, Sachs and Laudeho (427) report that P. blanchardi is a serious pest in Adrar (Mauritania), where studies on its control by Pharoscymnus ovoideus, Sec. and by C. palmarum are being undertaken.

When a heavy infestation of scale is found, generally on a weak shoot, cutting the affected fronds off low down and burning them is a simple and effective treatment. Martin (267) in the Fezzan found that diazinon killed this scale.

In Israel, Parlatoria is a serious pest. Although modern insecticides will now eradicate this scale so that the drastic torching the Californian dates received 50 years ago is no longer necessary, the Ministry of Agriculture is, except in special cases, not allowing spraying, because this also kills its predators. Predators have been bred and released in various places (100 000 for example at Yotvata but the number hitherto released has been too small to control the pest)./

Red Date Scale

The red date scale, *Phoenicococcus marlatti*, is probably found wherever the date palm is grown. Although at one time it was considered an injurious insect in the United States, Stickney *et al* (413) concluded that it was seldom damaging. They felt that this insect might contribute to general unthriftiness of some palms that had unusually heavy infestations, by causing drying of fruits on the palm and premature senescence of some older leaves. This scale insect is regarded as unimportant in the Old World.

The red date scale normally stays out of the light and is found massed on the white tissues at the bases of the leaves and fruit stalks, where they are protected by fibre and other leaf bases. Stickney *et al* (413) provided a comprehensive account of the biology of this insect. Under conditions of very heavy infestations, the underlying tissues were damaged to a depth of a few millimetres. The most effective control measure was to subject infested offshoots to a temperature of 50°C (120°F) for 65 hours in an insulated room. Chemical control in the field has been unsatisfactory. A few general predators are found in the United States, and in North Africa *Pharoscymnus anchorago* (Fairmaire) is an active predator.

Other Scale Insects

Asterolecanium phoenicis occurs on date palms in Iraq, Iran and Israel (72) and Saudi Arabia (329). It may cause severe injury by feeding on all parts of the leaves and on the fruit. Infested parts of the pinnae turn yellow and die. Heavily infested fruits are scarred and reduced in value. Kehat *et al*(228A) recommended control by spraying with malathion, azinphosmethyl, or diazinon combined with oil for control of nymphs in autumn and winter. In case of severe infestations, a summer spray may be applied to kill first instar scales before the crawlers settle on the fruit. Infestations by several other species of scale insects have been reported (250, 268, 329, 343, 424).

Date Bug

The date bug, *Asarcopus palmarum*, was first noticed in date plantings in the United States in 1922 (413), and has been reported from one locality in the western part of the Arab Republic of Egypt. These bugs sometimes weaken young palms but generally do not affect mature or healthy young palms. Date bugs usually feed on the white tissues of leaf bases, on the unexpanded parts of terminal leaves, and on the bases of the fruit stalks.

Adult female and male.
Dubas bug.

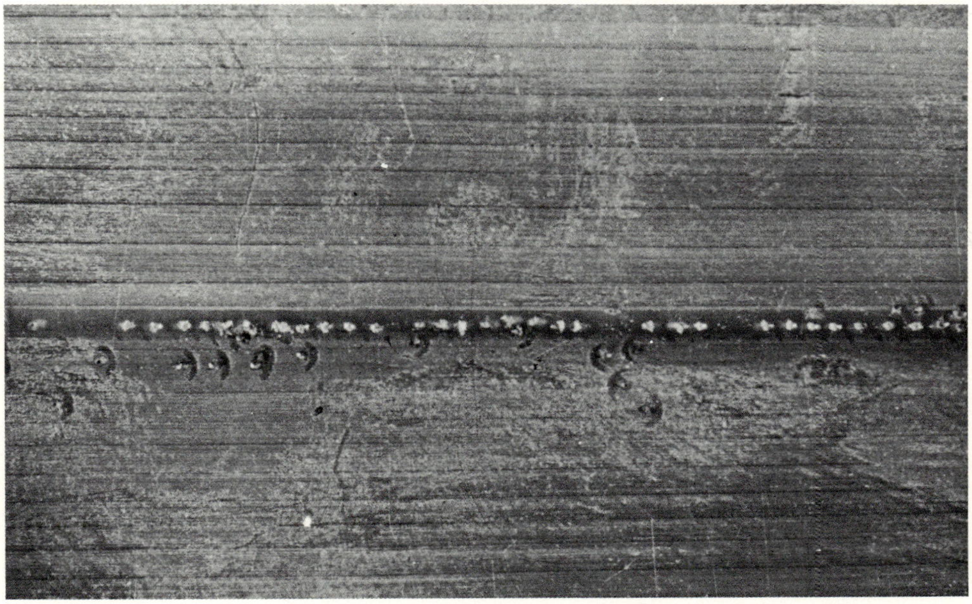

Dubas bug. Eggs laid in
the midrib of the leaflet.

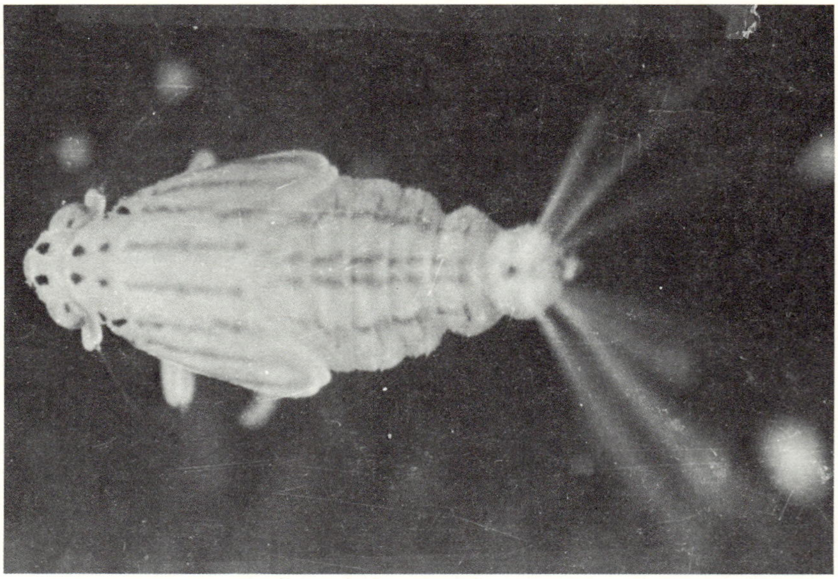

Dubas bug. Nymph.

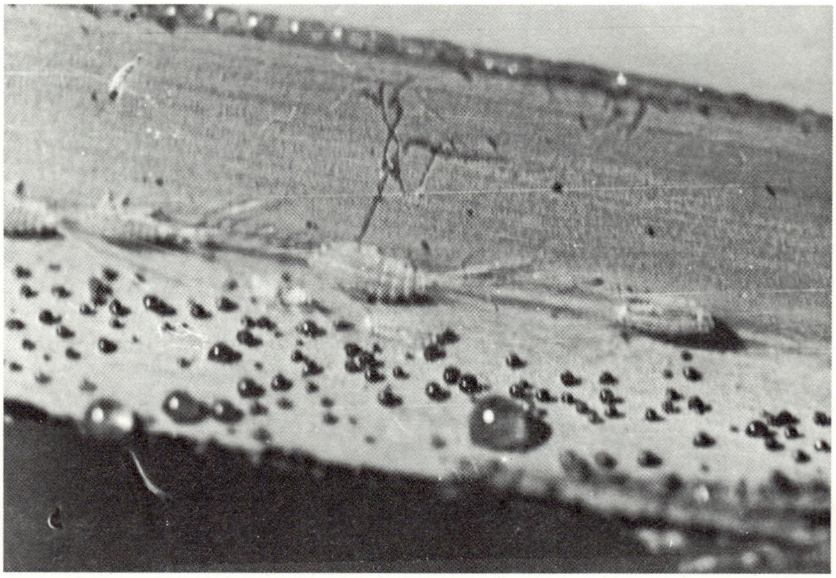

Dubas bug. Nymphs and
droplets of honey dew.

The eggs of the dubas bug are parasitized by a small chalcidoid. Larvae of the lace wing, Chrysopa carnes (Stephens), and adults of Coccinella septempunctata Linnaeus, C. undecimpunctata Linnaeus and Chilocorus bipustulatus Linnaeus prey upon the nymphs and adults (207).

The injury consists of superficial, irregular brown areas of damaged tissues. At times date bugs become numerous enough on small weak palms to cause the terminal leaves to droop, but control of this pest is not necessary in the United States except in a few isolated instances. Date bugs are active throughout the year, though less so in the winter. Populations on date palms consist primarily of nymphal stages. The bugs excrete honeydew which attracts ants.

Dubas Bug

Distribution and Economic Importance

The dubas bug, Ommatissus binotatus var. lybicus, also called the Old World date bug, is a serious pest of the date palm in Iraq (123, 161, 207, 243) and in Iran (128). It is found also in Oman and the Arab Republic of Egypt. If unchecked, the dubas bug may periodically develop large populations that ruin the date crop. No quantitative estimates of damage are available but Dowson(123), reported extensive damage among the seven million palms (approximately 28 000 ha) in the Basrah area of Iraq in 1934 and very severe damage in an area of about 800 ha. In 1935, the loss was estimated at US$400 000.

Injury

The first infestations occur about May, when nymphs begin to feed on the pinnae. Honeydew excreted by the insects and accumulated on the pinnae may cause them to become dry and discoloured. Honeydew accumulations on fruits may cause them to shrivel. Heavy infestations in which thousands of dubas bugs feed on each young leaf are readily observed because of the black drops and crusts formed as dust and fungi collect in the honeydew. Repeated heavy infestations cause weakening and death of some palms (123, 207).

Description and Life History

The adult female is 5 to 6 mm long and yellowish green. Two black spots are found at the base of the fronds and two on the pronotum. Often there are two or more spots on the vertex and one on each side of the seventh and eighth abdominal segments. The adult male is 3 to 3.5 mm long and lacks the four spots on the abdominal segments. The eggs are bright green changing to yellow; they are laid on all green parts of the palm except the fruits, with most deposited on the upper surface of the pinnae. Two generations occur each year in Iraq (207).

Control

Among insecticides used for control of the dubas bug, DDT and malathion were effective (207,243). Combinations of malathion with heptachlor and of DDT with diazinon were also effective. Malathion sprays gave good control in Iran (128) when applied early and thoroughly. Dichlorvos (DDVP) has been used successfully in low-pressure sprays or in aerial treatments at a rate of 4 l of product (48 percent active ingredient) per ha (161, 243).

Apache Cicada

Distribution and Economic Importance

The adult Apache cicada, *Diceroprocta apache*, may be found in large numbers during summer months in the desert regions of the southwestern United States. They infest dates sporadically and sometimes cause severe damage in individual palm plantings. During 1961 a considerable loss in date tonnage in California was attributed directly to oviposition by the female cicada (144). Comparable losses recurred in 1962 and 1963 and in most instances resulted from heavy migrations of cicadas from adjacent plantings of asparagus, the roots of which are a preferred host of immature stages of this cicada.

Injury

The primary damage to date plams is the loss of crop caused by female cicadas laying their eggs in the date strands. The injury is easily recognized as a typical wedge-shaped oviposition puncture; each series of punctures is made upward from the terminal date to the base of the strand. On severely damaged strands the dates shrivel, dry and fall. Secondary damage to palms may result from immature cicadas feeding on the roots.

Description and Life History

The adult of both sexes is stout, broad across the anterior end and is from 3.5 to 4.1 cm long. The general colour is brown, light to dark brown on the dorsal side and lighter brown on the remainder of the body, with a prominent light brown band across the back at the base of the transparent wings, which have brown-coloured veins. The nymphs are brown-coloured, thick-set and have strongly developed front legs fitted for burrowing in the ground. This cicada is believed to require two years for development in the ground, where it feeds on the roots of many different plants.

Control

No insecticidal control is feasible on dates because of the flight habits of the female. Paper covers, which in California are normally placed over ripening date bunches to protect them from rain, birds and certain insects, were used earlier than normal during 1962 for protection against the Apache cicada (144). Complete protection from ovipositing was obtained in all cases where covers were used early enough. Uncovered control bunches had from 25 to 80 percent loss of dates. It would be possible, however, for females to oviposit so early in the season that paper covers installed at such an early date might cause more damage than the cicada. Checking may occur on dates that are covered while still green; this results from increased humidity and sunburn due to intense heat under the paper covers.

Nitidulid Beetles

Distribution and Economic Importance

Four species of nitidulid beetles are of economic importance on dates in the Coachella Valley of California (258). They are the corn sap beetle, *Carpophilus dimidiatus*; the dried fruit beetle, *C. hemipterus*, also called fig beetle or two-spotted beetle; the pineapple beetle, *Urophorus humeralis*; and the yellowish nitidulid, *Haptoncus luteolus*. Dates on untreated palms in commercial growing districts may be completely ruined by these beetles. Although several species of nitidulid beetles are worldwide pests on dates, there are few estimates of damage from Old World date-growing areas, where *C. hemipterus* is the usual species (1,227,267,268,424). This beetle seems to show preference for certain varieties (268). In 1959, 30 percent of Bikrai fruit was damaged by *C. hemipterus* in Tripoli.

Injury

Nitidulid beetles damage ripening and curing dates on the palms, on the ground, and in storage by entering the fruits, usually at the calyx end, and feeding on the pulp. In addition to the primary damage caused by feeding, the unsightly and unpalatable frass that accumulates and the presence of larvae, the fruits are also subject to extensive damage by yeasts, other fungi and bacteria introduced by the beetles. Extensive spoilage may follow a heavy infestation, especially under humid conditions.

Description and Life History

C. dimidiatus is oblong, 2 to 3.5 mm long, slightly shiny, and varies in colour from black with a reddish tinge to brownish yellow, with the elytra always paler. C. hemipterus is oblong, 2 to 4 mm long with blackish buff spots on the elytra. It has been reported from the United States and the Old World. This species is very similar to C. dimidiatus in appearance. U. humeralis is oval to rather broadly oblong, with a length of 3.3 to 4 mm and a width of 1.6 to 1.9 mm. The upper surface of this beetle is very shiny and polished and the colour is brown to black, usually with a small pale spot within the humerus of each elytron. This, the largest of the four species described here, is generally distributed throughout the world. H. luteolus is oval to oblong, 2 to 2.5 mm long, moderately shiny and of a uniformly brownish yellow colour. This species is the smallest of the four and probably is confined to the United States. Several other species of nitidulid beetles not listed here are occasionally found feeding on dates in storage.

These nitidulid beetle species develop best at daily mean temperatures averaging around 32°C (90°F), which coincide with daily mean temperatures found in the Coachella Valley of California during June, July, August and September. High humidity at this time appears to be very important for the development of large populations of beetles. The first new source of food in the date orchards is provided during June when many green fruits drop (227). Fallen dates are infested by the first beetles of the season and develop the populations that later infest dates on the palms. Over a period of several months a single female may produce 500 to 1 000 eggs. The larvae pupate in the soil and overwinter in this manner, although adults can be found in almost any month feeding on fallen dates or on other decaying vegetable matter such as fruit and truck crop refuse. Lindgren and Vincent (258) discussed marked variations in the composition of beetle populations in the Coachella Valley. During the 1920's and 1930's, C. hemipterus was the dominant species in fruit picked from bunches, but by 1947, C. dimidiatus had assumed that role. During the five-year period 1947-1951 inclusive, C. dimidiatus made up 90 percent of the population in four years and 75 percent in 1949 only, when C. hemipterus increased from about 5 to 22 percent; the following year, C. hemipterus declined to less than one percent. The other species, U. humeralis and H. luteolus, averaged less than 2 percent annually in fruit picked from bunches during the same five-year period.

Control

Lindgren and Vincent (258) determined the initial and residual toxicity of many insecticides to nitidulid beetles. Malathion was selected for intensive testing as it was effective and less toxic to warm-blooded animals than parathion and dieldrin, the other effective insecticides. In 1952, when beetle infestations were unusually high, a 5 percent malathion dust reduced the number of live beetles in the fruit bunches from over 9 000 to 60. No live beetles were found in treated bunches for two weeks following application of malathion dust, and the bunches tended to remain relatively free from live beetles for up to eight weeks. Since 1953, malathion dust has been recommended for beetle control although other materials have been tested (143). In the Arab Republic of Egypt (1) malathion and carbaryl controlled these pests. Spraying with malathion emulsion was effective and left no residue in Israeli trials (227); it was preferred to 25 percent wettable powder. With the advent in the United States of mechanically harvested dates, early control of these beetles with malathion has become imperative because intact bunches of ripe fruit are left on palms for a long time (148). Nitidulid beetles in stored dates may be controlled by fumigation (258).

Beetles in Stored Dates

The rusty grain beetle, *Cryptolestes ferrugineus*, the saw-toothed grain beetle, *Oryzaephilus surinamensis*, and the merchant grain beetle, *O. mercator*, feed on dates in storage and cause fruit spoilage in much the same manner as the nitidulids. Species of *Oryzaephilus* occur in date-growing areas of the Old World and are commonly reported as minor pests. They are typical packinghouse and storage insects and are seldom found in newly picked dates. Fumigation, sanitation in the packinghouse, and high humidity keep these three pests under control.

Date Stone Beetle

The date stone beetle, *Coccotrypes dactyliperda*, occurs in the United States and Mexico (413), Algeria (424), the coastal areas of Libya (268) the Arab Republic of Egypt (71) and Israel (227). Damage is usually negligible, although substantial losses have been reported: 12 percent in the Aglawy variety (71), 45 percent in Bikrai (268) and more than 25 percent in Hadrawi (227).

The female beetle penetrates unripe fruit and lays its eggs on the seed. The larvae penetrate the seed and further generations may develop within the fruit. Buxton (71) reported finding single seeds with a total of 40 eggs, larvae, pupae and imagos. In Israel (227) spraying with one percent malathion (25 percent wettable powder) during July gave good control. In the United States only fallen fruits are attacked and the pest is unimportant.

Fig Beetle

The fig beetle, *Cotinis texana*, is found on dates in the United States. A few instances of serious damage to dates have been recorded during years of above normal humidity. Fermenting fruit is preferred by these beetles but, like the nitidulid beetles, they will puncture and enter uninjured dates. Nichol (308) gave a detailed account of the fig beetle.

Coleopterous Borers

Several species of beetles damage palms by boring into the leaves and trunks. *Pseudophilus testaceus* is a borer of the leaves and trunks of date palms. Infestations of 79 percent of palm trunks have been found in Iraq (207A). A mass of dark brown gum marks the entrance holes of larvae. This insect has one generation per year. The presence or absence of this borer and the number of emergence holes are major factors in determining the sale price of date plantings in parts of Iraq.

Oryctes elegans sometimes occurs as a serious pest in Iraq(207A) and Iran(128) The adults mine the surface of fruit stalks and rachises which may break if the damage is severe. A few strands or all of them may be affected, resulting in poorly developed fruit and heavy cullage. In Iraq not more than 2 percent of the fruit stalks are attacked severely enough to break and be a complete loss. The larvae are usually found inside the stems of dying or recently dead palms. There is one generation per year. No recommendations for chemical control measures are available.

Adults of *Phonopate frontalis* may mine the rachises of date palm leaves (207A). This borer also occurs in Libya (268) under the name of *P. frontalis uncinata* Karsch.

Apate monachus is a pest of date palms in Tunisia (110), Libya (268) and elsewhere in North Africa (72). The larvae bore into the midribs of date leaves and may make one or more galleries up to 12 cm long and six to eight mm wide, with occasional side galleries. The attacked leaves are subject to damage by wind and often break partially or completely. Although it is a minor pest, Pagliano (343) urged that broken leaves be pruned back to the trunk and that the affected leaves be burned immediately to destroy the insects. Calcat (72) reported that this insect also bores into trunks of weakened palms. He recommended control with paradichlorobenzene or carbon disulphide introduced into the galleries as fumigants.

Indian Palm Weevil

The Indian palm weevil, *Rhynchophorus ferrugineus*, is a pest of date palms in the Punjab regions of India and Pakistan (196,277). The beetles oviposit in the young leaf bases of the palm or in other succulent tissues. The larvae and beetles may feed in the crown of the palm or attack young palms at the places where offshoots have been removed. When the crown is attacked the larvae burrow down to the bud. As the insects mature and propagate, both immature and adult insects feed on and destroy the crown of the palm. Larvae and adults bore into trunks causing severely damaged palms to collapse. Injection of insecticides into burrows and painting tar on the wounds made by leaf pruning or offshoot removal are suggested control measures.

Indian Meal Moth

Distribution and Economic Importance

The Indian-meal moth, *Plodia interpunctella*, has been a pest on date palms since they were first planted in the United States. Although worldwide in distribution, it has been mentioned only as a pest of stored dates in Algeria (424), Libya (268) and Israel (117). The larvae of this moth are commonly associated with those of the raisin moth, and damage of up to 8 percent of Medjool dates has been reported in the United States (257).

Injury

The larvae of this species feed on ripe dates in the bunches, in the packing-house and on the ground. They enter through any break in the surface of the fruit, or through the calyx end of the fruit, or bore into intact fruits. They seem to prefer the drier, more mature dates in the bunches and even attack the hard date seeds. As the larvae feed they spin a web in which masses of frass are retained, and a web is also formed over the entrance hole. This damage renders fruit unsaleable.

Description and Life History

The short-lived adult moth is about 1 cm long. The outer half of the fore-wing is reddish brown with dark marking and the inner part is grey with a copper-coloured band separating the two areas. The life cycle varies from 36 days in summer to five months in winter. Eggs are deposited on the skin of the dates and average 170 per female moth. The dull-white larvae reach a length of about 13 mm. When mature, they leave the fruit to pupate among dates in the bunches, in the palm trunks, or in picking boxes. The pupae are light to dark brown and about 9 mm long.

Control

The Indian-meal moth was formerly a packing house control problem and fumigation of all newly harvested dates was mandatory. During the past few years more effort has been made to reduce populations in the bunches on the palm.

A 5 percent malathion dust applied about three weeks prior to the first fruit pick is recommended for combined control of this moth, the raisin moth and nitidulid beetles.

Raisin Moth

Distribution and Economic Importance

The raisin moth, *Ephestia figulilella*, infests fruit in the bunches and harvested fruit in the United States, and has also been reported from the Old World, where it is chiefly a pest of stored dates (1,95, 117). Larvae of this moth and of the Indian-meal moth have caused up to eight percent loss of Medjool dates in the United States (257).

Injury

Larvae of the raisin moth feed in ripening dates and, if the fruit is not fumigated, develop into adults. In addition to rendering the fruit unsightly and unpalatable, the larvae open the way for secondary organisms to invade and cause spoilage. This species and the Indian-meal moth account for a large percentage of the so-called hidden culls which cause a reduction in fruit grade in the packinghouse.

Description and Life History

Donohoe *et al* (118)reported on the biology of the raisin moth. The adult is about one cm long and grey in colour, with a few obscure dark bands and spots on the forewings. It lives about two weeks in warm weather; during this time the

Adult lesser date moth,
Batrachedra amydraula

Larvae, lesser date moth infesting
dates in the "Chimri" stage

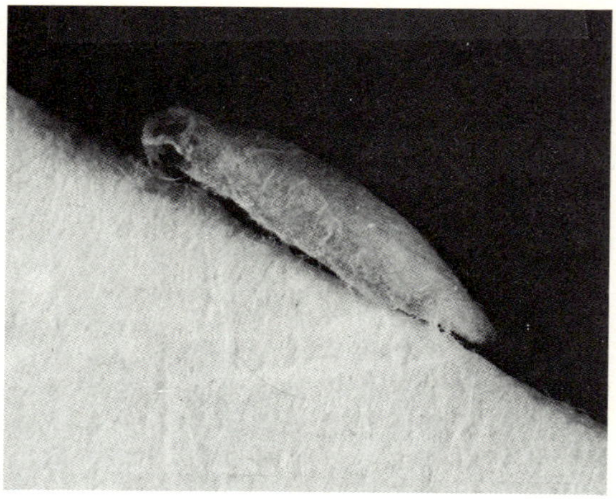

Pupa in a cocoon.
Lesser Date Moth.

Lesser Date Moth infesting
dates in the "Hababok" stage.

females lay an average of 350 eggs with a maximum of 690 being recorded. Eggs are laid on the surface of the dates and hatch in about four days. The larvae closely resemble those of the Indian-meal moth, but instead of being uniformly white the body is streaked with six rows of lavender-coloured dots. Larvae feed for about one month and, when full grown, are about 16 mm long. The pupae are brown in colour and enclosed in a silken cocoon which is spun in hiding places on the palm, or in the topsoil, or in storage buildings under boards and boxes, or in convenient cracks. Adult moths are most active in the early evening and remain in shaded protected areas during the daytime.

Control

A 5 percent malathion dust is used in a combined treatment for control of raisin moth, Indian-meal moth and nitidulid beetles. This treatment has kept the percentage of infested fruit at a low level. In the Arab Republic of Egypt (1), carbaryl and malathion effectively control this pest. In storage, fumigation with methyl bromide gas kills all stages of the raisin moth.

Ephestia Moths

In addition to the raisin moth, other species of *Ephestia* whose larvae damage dates during and after harvest, include the almond moth, *E. cautella*; tobacco moth, *E. elutella*; currant moth, *E. calidella*; and *E. dowsoniella* (62, 71, 462). *E. cautella* and *E. calidella* appear to be the more important species and damage stored dates in many countries including Iraq, Libya, the Arab Republic of Egypt and Morocco. Péreau-Leroy (351) reported that in southern Morocco, fumigation was not used and losses from *E. cautella* and other species were sometimes ruinous in the field and in storage.

Lesser Date Moth

Distribution and Economic Importance

The lesser date moth, *Batrachedra amydraula*, causes serious damage in some oases of the Fezzan in Libya (268) but in coastal areas the damage is slight. This moth is a serious pest in the southern and central plains of Iraq and in Iran(129, 161,243) causing many infested dates to be discarded at harvest.

Injury

The lesser date moth is a pest on immature dates in the bunch and in storage. According to Wiltshire (462), larvae may attack and damage large numbers of young fruits from April to June or July. Injured fruits usually fall, although the stout webbing spun by the larvae attaches some of the fruits to the strands or to other fruits before perforation. The larva enters the fruit by making a small hole near the calyx and then feeds on the pulp and immature seed. A larva seldom eats more than a third of the fruit before it seeks another one and may damage three or four fruits during its lifetime. Ramchandra Rao (368) found that about 30 percent of the fallen fruit had been attacked by this insect under presumably typical conditions in Iraq. The larvae that attack fruit just before ripening leave frass and other debris, so that affected fruit must be culled.

Description and Life History

The adult moth has a wingspread of about 11 to 14 mm and is of a brownish colour, marked by a grey longitudinal central stripe on the narrow forewings. The eggs are small, about 0.7 mm in diameter and yellow with a pearly sheen. The larvae, when full grown, are about 12 mm long. Some of them develop directly into adult moths, so that two or three generations occur in Iraq during the summer, and Martin(268) believes that even more generations may occur in Libya. The larvae of the last generation overwinter in the cocoon before pupating (368). Martin (268) found in Libya that only the last generation larvae developed normally when reared in captivity.

Control

Laville (243) and others(161) recommend two sprays per season of wettable DDT (50 percent wettable powder) at about five parts per 1 000. The first spray is applied one week after pollination and the second about 30 days later. Malathion and diazinon have also been tested.

Greater Date Moth

The greater date moth, *Arenipses sabella*, occurs in Iraq, Iran and Algeria, where it is a pest of dates on the palm and of stored fruit. In Iraq, 50 percent of the spathes and fruit bunches on 70 percent of the palms in some localities may be attacked (207A). Wiltshire (462) and Hussain(207A) reported that the first damage is seen in March when young larvae are found eating the tips of unopened spathes. When the spathes open the larvae enter and may strip the flowers and young fruits from whole strands. Their depredations are marked by a coarse silken tunnel which becomes littered with frass and plant fragments, and which may be 35 cm long when the larva is full grown. Larvae may bore into the base of fruit stalks and mine a gallery five to eight cm long in which bacteria and fungi cause secondary damage. Severely attacked bunches wither and the fruits shrivel but do not fall. Larvae also bore into young leaf midribs and feed on tender leaflets and immature dates. Adult moths are most commonly seen in April but are found from March to October. Hussain(207A)believes that two generations a year are produced. No recommendations on control are available.

Myelois Moths

Distribution and Economic Importance

Moths of the genus *Myelois* are pests in most of the date-growing areas of North Africa. All are considered by Lepigre(251) to be *Myelois ceratoniae* and include *M. ceratoniella* Fischer Edler von Roslerstamm, *M. pryerella* Vaughan, *M. zelleriella* Sorhagen, *M. phoenicis* Durrant and *M. decolor* Zeller. This view differs from that of Real (374) and some other authors. In Algeria, losses of 10 to 30 percent may occur (251). Calcat (72) reported that losses in North Africa from infestations by lepidopterous larvae, primarily those of *M. decolor*, amounted to 5 to 10 percent of harvested dates and might exceed 12 percent. Losses in the same range were reported by Wertheimer (456), but with more severe infestations.

Injury

The larvae attack maturing fruits, especially the drier ones, and cause damage by feeding, accumulations of frass and by the presence of live or dead larvae. If fruit is not fumigated promptly and stored properly serious losses may occur.

Description and Life History

Wertheimer (456) and Lepigre(251) have published the most recent studies on *Myelois* moths on date palms. Adult moths have a wingspread of 22 to 24 mm and are creamy white to grey or brownish, and may even be dark brown. Eggs hatch in three to seven days and the larval stage may last from one to eight months. Little is known of the pupal stage. The adult lives only three to five days, during which time the female may lay 60 to 120 eggs; three or four generations are produced annually, in the approximate sequence of April-May, June-July, August-November and November-March. Although the insects overwinter as larvae in various places, according to Lepigre(251) the moths that breed during spring and summer in dates left in the crown of the palms during the previous harvest are a primary source of reinfestation of maturing dates. Calcat (72) noted that only the last two generations of moths develop on dates in the bunch.

Control

In recent years, control of *Myelois* species by preharvest treatments has reduced the loss of fruit damaged prior to fumigation. Calcat (72) reported that benzene hexachloride applied at the beginning of the ripening period reduced infestations to about one percent. Wertheimer (456) and Lepigre(251) applied 10 percent DDT dust in various schedules from June to October and obtained good control, although not equal to that reported by Calcat (72). These treatments should be supplemented by fumigation with methyl bromide and careful storage of the fruit. *Myelois* larvae are aggressively cannibalistic and destroy great numbers of their own kind. The microhymenopterous predators of the larvae are unimportant.

Desert Locust

The desert locust, *Schistocerca gregaria*, occurs throughout the Old World date-growing areas (71, 72, 90, 351, 368 and 440). Heavy migrations into palm plantings are sporadic but may be devastating. The locusts eat leaves and fruits and may destroy an entire crop; they eat their own weight in food daily, and swarms are usually measured in terms of square miles. Thus, a swarm of 50 square miles represents about 10 000 tons of locusts and a comparable daily loss of crop and other vegetation as a result of their feeding. During a two-week period in 1954, locust swarms totalled about 1 000 square miles in the Sous Valley of Morocco and caused extensive damage to orchard and other crops. The damage was estimated at $3 million (367). During the winter of 1958-59 an invasion of this locust in Israel lasted 14 days (90). Palms were completely defoliated and did not renew growth until the following April. Heavy defoliation of palms is undoubtedly reflected in reduced crops for several years because renewal of a full crown of leaves requires at least three years under optimal growing conditions.

Control

Chemical control is effective if applied properly and timed to kill the locusts before they attack the palms. The coordinated use of baits and dusts in breeding and swarming areas and the use of aerial spraying on both ground and flying swarms of locusts has been successful. Since 1959, aerial spraying has come into widespread use. Examples of kills obtained by spraying locusts from aircraft, in terms of locusts killed per gallon of spray, were: 20 percent dieldrin, 2 million; 30 percent dieldrin, 3 million; 11 percent lindane, 1 million (138); 85 percent diazinon, 3 million (367). The numerous factors that affect the success of spraying, baiting and dusting are summarized in various reports (113, 138,139, 140, 141).

Hymenoptera

Many different species of bees, hornets and wasps feed on all kinds of fresh fruits and quite often members of this group cause some damage to dates, particularly soft-fruited varieties. This has occasionally led growers to use various types of cloth covers over the ripening fruit to protect it against these pests as well as against rain and birds (43).

Termites

In Mauritania, Lefèvre (246)reported that termites of the genera Bellicositermes, Ibostoma, Coptotermes and Psammotermes damaged date palms. Control of termites on individual palms was considered more practical than trying to eliminate the subterranean galleries in which most of these species breed. One litre of a 5 percent mixture of benzene hexachloride (25 percent active material) in water applied around the base of each palm provided protection from termite attack for several months.

Another termite, Microcerotermes diversus, is a pest of date palms in Iraq (207A).This termite attacks roots, trunks and leaves. It makes galleries in the trunks of weak palms and may cause them to collapse.

Aphids

Date aphids, not identified in text(161), have become widespread in Iraq during the past 45 years and cause substantial reduction in quality and loss in yield of dates. Malathion at 240 g per 100 gallons of water is recommended for spraying. Dichlorvos (DDVP) is a potentially suitable insecticide for aerial spraying of this insect.

3. FUMIGATION AND PACKINGHOUSE PEST CONTROL

Fumigation of all newly picked dates, as they are brought into the packinghouse, arrests insect infestations that occur within the fruit. Fumigation also helps prevent reinfestation within the packinghouse before the fruit reaches the consumer. Walker and Mitchell (450) describe techniques for fumigating dates and the types of chambers and equipment employed in packinghouses in California. Methyl bromide is now used almost exclusively in the United States, although carbon disulphide, hydrogen cyanide gas and a mixture of ethylene oxide and carbon dioxide had been applied previously.

Armitage and Steinweden(22) in the United States and Junaid and Nasir (222) in Pakistan have reported on the use of methyl bromide as a date fumigant. Brown and Heuser (63) found that methyl bromide penetrated boxes of compressed dates more effectively at reduced pressure than at atmospheric pressure. Methyl bromide is a particularly satisfactory fumigant for date insects because of its high toxicity to all stages of insect life. Its apparently low chemical reactivity with date tissue and low sorption lessens residual odours and flavour changes in the fruit. Methyl bromide is highly toxic to man and should be used with safety precautions. California date packers at the present time use one lb of methyl bromide per 1 000 cubic ft of storage space at temperatures between 16°C (60°F) and 32°C (90°F) for four hours. Recommendations call for a 24-hour exposure time, but apparently four hours are sufficient to kill all pests and to meet packinghouse requirements in the United States (147).

Recent recommendations in Iraq(161) call for 1.5 lb of methyl bromide per 1 000 cubic ft. Outdoor fumigation of dates in field boxes covered with polyethylene tarpaulins at the rate of one lb methyl bromide per 1 000 cubic ft was reported in 1966 (161). A similar method of fumigation is used in California. Outdoor fumigation has been investigated in Israel (73).

Spraying of packinghouses before the date-packing season with malathion at the rate of 0.5 - 1 g active ingredient per square metre has been recommended.

4. ROOT-KNOT NEMATODES

Root-knot nematodes were found in the Coachella Valley of California on date palms in 1925 (438) and are now known to be widely distributed there in commercial date plantings. Buhrer *et al* (65) first reported the occurrence of root-knot nematodes on this host. Jensen(219) reported *M. incognita* (Kofoid and White, 1919) Chitwood, 1949, on roots of date palms in western nurseries. Authors after

1950 identified the nematodes as *Meloidogyne* sp. Carpenter (75) reported that root-knot nematodes, principally *M. javanica* (Treub, 1885) Chitwood, 1949, can severely damage or kill date palm seedlings. Young seedlings of 50 date varieties were susceptible to infection by root-knot nematodes; more than 90 percent of the seedlings were killed prior to emergence when seeds were sown in heavily infested soil. Secondary damage by fungi to roots of infested field-grown palms seems to be an important factor in the deterioration and death of roots. Minz *et al* (279) reported on occurrence and control of root-knot nematodes on date palms in Israel.

LIST OF MITE AND INSECT PESTS DISCUSSED ABOVE

Scientific name and author	Common name	Order-Family
Apate monachus Fabricius		Coleoptera-Bostrichidae
Arenipses sabella (Hampson)	Greater date moth	Lepidoptera-Pyralidae
Asarcopus palmarum Horvath	*Datebug	Hemiptera-Issidae
Asterolecanium phoenicis (Ramchandra Rao)		Hemiptera-Asterolecaniidae
Batrachedra amydraula (Meyrick)	Lesser date moth	Lepidoptera-Cosmopterygidae
Carpophilus dimidiatus (Fabricius)	*Corn sap beetle	Coleoptera-Nitidulidae
Carcophilus hemipterus (Linnaeus)	*Dried-fruit beetle	Coleoptera-Nitidulidae
Coccotrypes dactyliperda (Fabricius)	Date stone beetle	Coleoptera-Scolytidae
Cotinis texana Casey	Fig beetle	Coleoptera-Scarabaeidae
Cryptolestes ferrugineus (Stephens)	*Rusty grain beetle	Coleoptera-Cucujidae
Diceroprocta apache (Davis)	Apache cicada	Hemiptera-Cicadidae
Ephestia calidella (Guenea)	Currant moth	Lepidoptera-Phycitidae
Ephestia cautella (Walker)	*Almond moth	Lepidoptera-Phycitidae
Ephestia dowsoniella (Richard & Thompson)		Lepidoptera-Phycitidae
Ephestia elutella (Hübner)	*Tobacco moth	Lepidoptera-Phycitidae
Ephestia figulilella Gregson	*Raisin moth	Lepidoptera-Phycitidae
Haptoncus luteolus (Erichson)	Yellowish nitidulid	Coleoptera-Nitidulidae
Microcerotermes diversus Silvestri		Isoptera-Termitidae
Myelois spp.		Lepidoptera-Phycitidae
Oligonychus pratensis (Banks)	*Banks grass mite	Trombidioformes-Tetranychidae
Oligonychus afrasiaticus (McGregor)	Date mite	Trombidioformes-Tetranychidae
Ommatissus binotatus var. lybicus Bergevin	Dubas bug	Homoptera-Tropiduchidae
Oryctes elegans Prell		Coleoptera-Scarabaeidae
Oryzaephilus mercator (Fauvel)	*Merchant grain beetle	Coleoptera-Cucujidae
Oryzaephilus surinamensis (Linnaeus)	*Saw-toothed grain beetle	Coleoptera-Cucujidae
Parlatoria blanchardi (Targioni-Tozzetti)	*Parlatoria date scale	Hemiptera-Diaspididae
Phoenicococcus marlatti Cockerell	*Red date scale	Hemiptera-Dactylopidae
Phonopate frontalis Fahraeus		Coleoptera-Bostrichidae
Plodia interpunctella (Hübner)	*Indian-meal moth	Lepidoptera-Phycitidae
Pseudophilus testaceus Gahan		Coleoptera-Cerambycidae
Rhynchophorus ferrugineus Olivier	Indian palm weevil	Coleoptera-Curculionidae
Schistocerca gregaria (Forskal)	Desert locust	Orthoptera-Acrididae (= Locustidae)
Urophorus humeralis (Fabricius)	Pineapple beetle	Coleoptera-Nitidulidae

* Common names that have been accepted by the Committee on Common Names of Insects of the Entomological Society of America (77). The other common names listed are those widely used.

5. MITE AND INSECT PESTS OF MINOR ECONOMIC IMPORTANCE

The following notes, collected by Dowson, indicate the species of mites and insects which have been of some local economic importance as pests of dates. They may prove to be of greater or lesser importance in the future.

ERIOPHYIDAE

Mackiella phoenicis (Kief) and Tumescoptes trachycarpi K.
Mohamed, I.I. and El-Haideri, H.S. (280) report these mites from Iraq where they infest the folded leaves causing stunting and deformation.

TENUIPALPIDAE

Tenuipalpis eriophoides, Baker
Pritchard and Baker (363) reported this mite from Basra (Iraq) from an unknown host. Mohamed, I.I. and El-Haideri, H.S. (280) have found it on date palms, mainly on the leaflets, where it causes some chlorosis. They consider it to be a minor pest.

TETRANYCHIDAE

Eutetranychus banksi, McGr. (= Anychus latus (Can) Sayed;
= A. orientalis (Zacher) Klein)
Martin (267) found this mite on dates at Sidi Khalifa on the Cyrenaican coast. This pest is of no economic importance.

COCCIDAE (Scale Insects)

Aleurocanthus bambusae, Peal.
Singh (407) reports this scale on the date palm in India.

Aonidiella orientalis, Newst. (= Aspidiotus cocotiphagus, Marlatt.)
Hauser (205), An. Ent. Soc. Amer., XI: 157, states it attacks the date palm.

Aspidiotus cocotiphagus, Marlatt. See Aonidiella orientalis, Newst.

Aspidiotus hederae, Vall., Oleander Scale.
Martin (267) found Oleander Scale on dates at Latrum on the Cyrenaican coast. See also Essig (151) who mentions the presence of this scale in California, and Afchar (2) who found it in Iran.

Asteroleacnium phoenicis, Ram. Rac
Following Imms (213), instead of Lepesme, transferred from Asterolecaniidae to Coccidae. In Dowson and Aten (129, p. 186), owing to a misprint, it is stated that it is a pest in Israel. For Israel, read Iraq. Unimportant in Iraq. Does not appear to have been reported from Israel.

Chrysomphalus dictyospermi, Morg. The Citrus Red Scale.
This scale is common on citrus in Basra but rare and unimportant on dates there.

Coccus hesperidum, L.
Lepesme (250) states that it has been found on date palms.

Duplachionaspis chaetachmae, Brain
Hall (193) found this scale on the date palm in Southern Rhodesia.

Fiorinia fioriniae, Targ.
Lepesme (250) states that the date palm is one of the numerous hosts of this scale.

Icerya aegyptiaca, Douglas.
Newstead (307) found this scale on a young date palm in Zanzibar. No other record. This scale is of no economic importance on dates.

Pseudaspidoproctus hyphaeniacus, Hall
Martin (267) reports that this scale was found, always in small numbers, on petioles and leaf bases in the oases of Hon, Sebha, Brak and Ubari in the Fezzan. It has very little economic significance.

Pseudococcus brevipes, Ckll.
Costa Lima (103) reports this scale on date palms in Egypt and Brazil and that it is preyed upon by Pseudiastata brasiliensis.

LEPIDOPTERA (Butterflies and Moths)

AGROTIDAE

Spodoptera mauritia, Boisd.
On the date palm in Hawaii (Pemberton, Haw. Plant Rec. XXXVII, 4:168). Unimportant.

BRASSOLIDIDAE

Brassolis astyra, God. (Formerly included in the Nymphalidae).
The larvae eat the fronds of date palms in Brazil and Argentina. (Lepesme, 250). Unimportant. Parasitized by Xanthozona melanopyga.

SCHRECKENSTEINIIDAE (Heliodinidae)

Stathmopoda trissorrhiza, Meyr.
Rahman and Abdul Wahid state that in India this moth attacks the dates on the palm and in store (366).

COLEOPTERA

Bostrychoplites zickeli, Mars.
Tropical Africa, Algeria and Tunisia. The larvae burrow in the dried fronds and trunk of the date palm (253, 469).

Enneadesmus trispinosus, Ol.
Lesne (253) states that the larvae burrow in the dry fronds of the date palm in North Africa. This beetle is preyed upon by Teretrius brunneus, Lew. (Lesne, 253) and Macrotelus pallidicolor, Frm. Lepesme (250) gives the distribution of T. brunneus as the Mediterranean, the Algerian and Tunisian Sahara, Lower Egypt and Iraq.

Phonapate frontalis, Fahr., sub-species arabis, Lesne
Lepesme (250) gives the habitat as Tripolitania, Red Sea, Gulf of Aden and Persian Gulf shores and states the Cumming found the larva of this beetle in a palm on the coast of the Persian Gulf.

P. frontalis, Fahr., sub-species uncinata, Karsch.
Lepesme (250) states on the authority of Dr. Chobaut and Dr. Vauloger that this beetle is found in North Africa and West Asia and that its larva bores into dried date fronds, tamarisk, and bamboo.

BUPRESTIDAE

Julodis caillaudi, Latr.
This large, green, irridescent, buprestid beetle was found in December 1947 eating the leaflets of date palms on the Ghaba Government Scheme and at Fargai (Sudan). It was called by the cultivators variously "Abu Zunaan",

"Abu Dinaan" and "Kudundar". Although the damage done to individual palms was considerable, the total damage caused by it in the province (the Northern) was negligeable.

CERAMBYCIDAE

Talhouk (N.D.) states that larvae of this family bore into dead or drying palms in northeast Arabia, but gives no specific, or even generic, name.

Apatophysis barbara, Lucas
The specific name is from Lepesme (250), who quotes from Surcouf (416). Surcouf states that the larvae burrow into the midribs of date palms in Algeria. Lepesme adds Tunisia.

CHRYSOMELIDAE

Himatidium gomes-costai, Bond.
Eats the leaflets of the date palms. Moderately important in Brazil. (Bondar, 46).

CRYPTOPHAGIDAE

Cryptophagus cellaris, Scop.
Lepesme, quoting F. Pierre, states that this beetle is found on the date palms around Touggourt, Algeria.

CUCUJIDAE

Laemophloeus ater, Oliv.
Martin (267) reports this beetle as a pest of stored dates in Libya but less common that Plodia interpunctella, Oryzaephilus surinamensis, Tribolium confusam and T. castaneum.

CURCULIONIDAE (Weevils)

Notesiaspis rubiginosus, Hust.
Bondar (46) reports this beetle as boring into the main axes of the fruit bunches and of the fronds of the date palm in Argentina.

<u>Rhynchophorus</u> <u>phoenicis</u>, F.
Lepesme (250) states that this beetle has habits similar to those of <u>R</u>. <u>ferrugineus</u>, Ol., and that it often attacks a palm after <u>Oryctes</u> has attacked it. He gives <u>P</u>. <u>dactylifera</u>, L., as one of its host plants, but does not state in which country or countries it is attacked by this beetle. He gives its habitat as the whole of tropical Africa, Abyssinia, and perhaps Mesopotamia. He adds that the larvae are preyed upon by the scaritid, <u>Niochryopus</u> <u>savagei</u>, Hope.

DERMESTIDAE

<u>Trogoderma</u> <u>versicolor</u>, Creutz.
(Donohaye and Calderon, 117) in stored dates in Israel.

ELATERIDAE (Click Beetles)

<u>Agrypnus</u>, a new species, near <u>luridus</u>, F.
This beetle was found in small numbers in northern Somalia in 1952 eating date-shoot fronds under the shelter of the wrapping.

MYCETOPHAGIDAE

<u>Typhaea</u> <u>stercorea</u>, L. (The Hairy Fungus Beetle)
Donahaye and Calderon (117) report it from stored dates in Israel.

NITIDULIDAE

<u>Carpophilus</u> <u>decipiens</u>, Horn. See Dowson and Aten (129)

<u>C</u>. <u>dimidiatus</u>, F. See Dowson and Aten (129)

<u>C</u>. <u>hemipterus</u>, L. (The Dried-Fruit Beetle)
The Dried-Fruit Beetle in Iraq does not attack dates on the palm, only dates on the ground and in store, but in Northern Somalia, it attacks dates not only on the ground but also on the palm if the dates have first been damaged by <u>Stalagmosoma</u> <u>cynanchi</u>, G. and D. Martin (267) states that in Libya this beetle (<u>C</u>. <u>hemipterus</u>) is attracted by injured or over-ripe dates on the palm and "even found to infest apparently sound fruit but only shortly before, or when the dates are ripening".

SCARABAEIDAE (Chafers, etc.)

Oryctes boas, F.
De Miré (112) states that this beetle is "extremely common in all tropical Africa and that it reaches in the Ennedi and doubtless in the Republic of Mali the date zone, where it is found with sahariensis, and may be expected to attack this palm as it does others".

O. monoceros, Ol.
Tropical Africa. Polyphagous. De Miré (112) states "it forms a very isolated colony in the date grove of Faya in Borkou, where its depredations are strictly confined to the date palm".

O. rhinoceros, L.
Kurian and Pillai (235, p. 20) state that this species is capable of attacking the date palm.

O. sahariensis, De Miré
De Miré (112, pp. 251-2) states that this species is found in all the date gardens of Borkou and of Ounianga as well as in the grove of Fada in the Ennedi. It is also found in the great valleys of Tibesti as well as on the northern slope (Bardagué; Zoumeri) and the south-west border (Zouarké, Maro, Tougouma) and at the foot of the Emi Koussi (Térouane) where the adult is in flight at nearly all seasons. Lastly, it is found also in the valley of the Nile in Upper Egypt.

O. sinaïcus, Wek (= desertorum, Arw.)
Has been found in Basra, where it has similar habits to elegans, but it is much less common than elegans there. Sir Guy Marshall (correspondence) adds that this Oryctes also occurs in northwest India and Iran. Arrow (25) gives the distribution as Aden (Lahej), Arabia, Iran, Masqat, Pakistan (Karachi).

Phyllognathus excavatus, Forst. (= silenus, F.)
Surcouf (417) states that in Algeria the adult beetle can be found in rotting dates and dust collected round the growing point of the date palm, and that the larva eats the living tissue.

Stalagmosoma cynanchi, G. and D.
This cetoniid beetle was a serious pest at Dubar, near Berbers in Somalia, in 1951, where it attacked the ripening dates in the early rutab soft stage.

Strategus julianus, Burm. (The Palm Rhinoceros Beetle)
Flooding the beetles out of their holes with water is the best control measure (Stickney, 413).

Xyloryctes jamaicensi; Drury
Nixon (333) states that in the lower Rio Grande Valley of Texas date offshoots are occasionally injured by this large, rhinoceros beetle, which bores into them just below the soil surface.

TENEBRIONIDAE (Meal Worms, etc.)

Tribolium castaneum, Hbst. (= ferrugineum, Auct.)
Found in Basra in 1937. Martin (267) reports it from stored dates in Libya.

HYMENOPTERA (Ants, Bees, Wasps, etc.)

VESPIDAE

Polistes annularis, L.

P. apachus, Sauss.

P. exclamans, Vier.
Nixon (333) reports that these three wasps often seriously damage date fruit in the district round Winter Haven, Texas. He states that the only known method of preventing damage is to enclose the bunches in cloth bags prior to ripening.

P. carapyta, Oliv.
Mello and Carrera (1959) report that this wasp occasionally attacks maturing dates in Argentina.

P. gallicus, L.
Hussein (209) reports this wasp as a pest of ripening dates in Iraq.

Vespa orientalis, F.
Bodenheimer (44) reported that Vespa orientalis, F. attacks dates in Palestine. Buxton (71) states that he believes it never attacks the dates on the palm in Iraq unless they have been previously pecked by birds. This is a pest of moderate importance in Iraq.
Hilgeman, who was in Israel in 1952, reported (201) that the worst pest of Israeli dates was V. orientalis. He recommended wire-net bags. Today, almost all of the crop is bagged, and wasp damage is now insignificant.

FORMICIODAE

In 1712 Engelbert Kaemfer gives an account of the pests of the date palm, and mentions that in the Yemen, large ants are brought down by the shepherds from the mountains in branches of trees, in holes in which the ants live. These branches are sold in shops and markets in Mocha, where they are bought by cultivators of date palms troubled by other ants and other pests. The new, large ants from the mountains, when released in the coastal palms, chase away the harmful ones.

TRYPETIDAE (Fruit Flies)

Ceratitis capitata, Wied. (The Mediterranean Fruit Fly)
Willar (Proc. Haw. Ent. Soc., VI:272, 1927) reports this fly in dates in Hawaii, and Bodenheimer (44) in Israel. Lepesme (250) lists 13 hymenopterus predators on this fly.

6. BIRDS 1/

Insects and mites are the most serious pests of dates and date palms, but the total damage done by birds throughout the date world must be immense.

Buxton (71) has the following note about birds in Iraq: Hypocolius ampelinus has been observed in Baghdad gorging itself and damaging many more dates than it could devour (see also Gilliard, E.T., 1962, p. 320). The bulbul (Pycnonotus leucotis mesopotamiae, Ticehurst) and the house-sparrow (Passer domesticus biblicus, Hartert) (Ar. zarzuur) (see also Husein, A.A., 209, Paper 19, p. 14, who has P. domesticus biblicus, Hartest) are destructive. Hypocolius ampelinus, Bonaparte, is the Grey Coly-shrike. The bulbul is certainly also a feeder on dates in Basra, and so are the common dove (hamaam) and the hooded crow (ghuraab).

1/ Contributed by V.H.W. Dowson

H.H. King informed Buxton that the Spanish sparrow (*Passer hispaniolensis*) pecks young dates in the Sudan. It is also a pest in Libya and was found to be somewhat serious at Ghadames in 1959.

Milne (277) states that parrots appear to be one of the chief enemies of dates in the Punjab.

Bacon (27), with reference to dates in the Sudan, states: "Birds, notably the gluttonous bulbul, will take the fruit of any isolated sweet varieties but are not very troublesome in the main producing areas". Parakeets come up from the south as far north as Bauga, about 80 km north of Atbara, where they have been reported to damage the date crop.

In places where birds are a serious pest, simple gunpowder bangers tied at widish intervals to a long length of slow-burning fuse suspended from a frond are useful and so is the more elaborate and expensive acetylene generator which explodes a mixture of the gas and air by means of an electric spark at intervals determined by the setting of the water feed drips.

In Somalia the crow (*tuka*), the bulbul (*degin*), the weaver bird, and the tick bird (*huuriyu*) all ate dates if they got the chance. The first is a serious pest, but the damage may lessen proportionately as the crop increases.

Nixon (328), writing of the date palms in the Baja Peninsula in Mexico, states: "Perhaps the most serious losses of fruit result from the depredations of birds, particularly the raven by day, and of the cocomixtle, a relative of the racoon, by night."

In the Punjab, Milne (277) states that the bird scarers in the date gardens, when the dates are softening, frequently hang bells and other contrivances for making a noise on the palms, which they shake occasionally by means of strings.

The usual way of keeping birds away from ripening dates, in parts of the Punjab, in Pakistani Baluchistan, and in Irani Baluchistan, and (for the very late Hilaali variety), in most of southern Iran, is by bagging the bunches in matting bags made of woven palm leaflets. The palm preferred as a supplier of leaflets in Baluchistan is the dwarf fan palm, *daaz* or *mazari*, *Nannorhops ritchieana*, H. Wendl., because its leaflets are longer and stronger than those of the date palm. The baskets, *sund* in Baluchi, are woven loosely, so that there is moderately good circulation of the air through the bunches. After harvest the *sund* may be used for the storage and the transport of the dates.

Somali women are skillful in making mats and other articles out of date palm leaflets, and learnt readily how to make mat bags in which to enclose the bunches on the palms in the Government Date Introduction Stations.

7. MAMMALS 1/

Migratory fruit bats in the Tabora and neighbouring districts (Tanzania) (where, in 1959, there were about 6 500 date palms), together with birds would destroy the whole date crop if it were not protected by sacking.

Hughes-Buller (206, p. 8) states that "the date palms in the ravines and tableland of Dhrun, at the northwest end of the main Makran range (Baluchistan) attract bears".

The cocomixtle, reported by Nixon (328) as one of the two most serious pests of dates in the Baja Peninsula of Mexico, has already been mentioned.

Hyenas were troublesome in some of the young groves in Israel, tearing through the wire-mesh bags on the low bunches and eating the ripening dates. They were caught in traps and now do not do much damage.

Jackals (dawa in Somalia) will eat dates on low palms, but the only places where it seems they did much damage were at the Date Introduction Stations of Barakallah, near Berbera, and at Takhushe, near Zeilah, in Somalia, and in some of the young plantations in Israel until they were fenced in.

Porcupines (*Hystrix cristata*, L. - Somali, Anaqub) were a nuisance at the Bulhar date introduction station in Somalia, eating through the young fronds of the date offshoots near their bases, until the wire fence round the station was completed. They also did similar damage at the Dubar station. In Israel they did considerable damage to newly planted shoots until they were controlled by trapping and fencing.

Popenhoe (356) states that: "Rats sometimes destroy part of the date crop, of which they seem to be particularly fond. A horde of migratory rats which visited the date orchard at Tempe, Arizona, a few years ago, caused great damage. At Baghdad, a squirrel-like rodent has the same habit, and the growers protect a choice tree from him by putting a collar of tin a foot wide around it, at a height of 3 ft from the ground. This would probably be an effective treatment here, and would be a permanent protection".

1/ Contributed by V.H.W. Dowson

The ground squirrel or tree rat Xerus rutilus dabagalla, Heugl. (Dagabale) is a minor pest of dates in Somalia.

The Short-Tailed Bandicoot Rat (Nesokia indica, Ar. jureidi, Pers. muush-i-sahraai) is a serious pest of dates in Khuzistan (Iran). It burrows underground and it eats the roots of young and old palms, killing young ones often quickly and older ones more slowly. In the Khuzistani date gardens on the Shatt-el-Arab and in the Iraqi gardens on the same river, the damage by this rat is much less than where irrigation is by pump, because the tidal irrigation provides the rat with less suitable conditions for its burrowing. What is probably the same animal is found in the date gardens on Qatif (Hasa, Saudi Arabia).

Effective control can be achieved by putting a little poisoned grain into the mouth of each burrow. The grain (wheat) is made sticky by being shaken up with a little cooking oil, and to about a kilo of the mixture is added a little (about a desert spoonful) of finely powdered zinc phosphide; and the whole is then thoroughly mixed.

Rattus rattus, probably R. r. frugivorus (Ar. Farr en-Nakhal, Pers. Muush-i-Khurmaa) does some damage to the ripening dates in Khuzistan.

Wild Pig (Ar. khanziir, Pers. Khuuk) do some damage in the date gardens along the Karun River in Khuzistan (Iran). They will destroy any bunches of dates on young palms that they can reach, but the greater damage is to the undercrops and in breaking down the thorn fences. It is said that the Irani Air Force, a few years ago, bombed the large tamarisk thicket by the river, in which the pigs used to sleep by day, and killed a large number of them; but, since then, their number has much increased again.

8. DISEASES

Bayoud

Introduction

Bayoud, or fusariose, is the most important disease of date palms. It is confined to Morocco and southwestern Algeria but is a threat to date palms everywhere. The causal organism is Fusarium oxysporum Schlect. var. albedinis (Killian and Maire) Malençon. Cipolla (89) reported studies of a fusariose on date palms in Argentina that might be related to bayoud. Péreau-Leroy (349, 350, 351) summarized the history of bayoud and previous investigations, including the extensive and valuable contributions of Malençon. Teisseire (424) emphasized means for controlling bayoud. In 1965, Toutain (430) reviewed the bayoud problem in Morocco and Algeria. Bulit et al (66) and others (16) reported recent investigations of bayoud.

Distribution

Bayoud was first observed in Morocco north of Zagora in the Draa Valley sometime before 1890. Since then it has spread continuously and, by 1956, the only date-growing area in Morocco still free from bayoud was at Marrakesh. It spread more rapidly toward the east than the west (430) and by 1898 was recognized at Figuig on the edge of the Algerian Sahara. Thereafter, bayoud appeared in several locations in Algeria south and west of El Golea; it appeared in date plantings at Metlili, north of El Golea, sometime between 1949 (430) and 1958 (72). Bulit et al (66) noted recent advances of the disease in Algeria and discussed the potential hazards of further spread.

Economic Importance

By 1950 bayoud had killed 10 million palms in Morocco (351), including the greater part of the vigorous and productive palms of the best commercial varieties, such as Medjool and Bou Feggous. The remaining five million palms were predominantly seedlings and varieties of only local interest. Areas that formerly had 300 to 400 palms per ha, and where intercrops were virtually unknown, such as at Bou Denib and at Haut R'teb, were reduced to five to 10 palms per ha. The losses are compounded by the fact that in many oases with an irregular surface water supply the loss of date palms removes the staple food crop and medium of exchange for which there are no adequate substitutes. Even in oases receiving a regular supply of surface water no other food crop gives yields comparable to those of the date palm. The ravages of bayoud have forced the population to emigrate from some severely affected areas.

Since the fine commercial Deglet Noor date is susceptible (72, 430), bayoud is a particular threat to the principal date-growing areas of Algeria and Tunisia because of intra-Saharan traffic among desert peoples.

Symptoms

Symptoms of bayoud appear first on one or more of the recently matured leaves. Typically, some spines or pinnae toward the base of one side of the leaf become white and then adjacent pinnae whiten as the disease progresses upward along one side of the rachis to the apex. After one side has been affected, the whitening and dying of pinnae progress down the opposite side of the leaf until it is killed. This may take a few days to several weeks. As the pinnae die, a brown stain appears on the dorsal side of the rachis and the surface of the diseased area becomes depressed. The stain and depression may extend the length and width of the rachis, except on the part hidden under the fibre which remains alive and turgid indefinitely. A transverse cut through the affected parts of the rachis reveals brownish red necrotic areas.

Concurrently, adjacent or opposite leaves are affected in like manner, until finally one or two whorls of leaves are killed. The old leaves wither naturally as on unaffected palms but no new leaves form and the terminal bud dies, leading to death of the palm. The average time from the appearance of symptoms to death is six months to two years; however, palms may die in one month or linger for ten years or more.

Sometimes the symptoms develop differently. A brownish lesion may appear on the back of the rachis and extend upward until the rachis becomes so narrow that all tissues are affected and the whole tip of the leaf whitens and dies. The other pinnae then die downward to the base. Other minor variations in the early symptoms may occur and diseased pinnae may be tinged brownish or reddish depending on the variety. Experienced observers may detect a general slight yellowing in some recently matured leaves one or two months before the appearance of typical symptoms.

When the trunk of an affected palm is split, vertical reddish bands of diseased tissue a few square cm in extent are found as well as individual discoloured vascular elements. Bulit et al (66) traced the fungus in diseased palms from the roots through the trunk to the leaf bases and to the terminal bud. The causal fungus was isolated from typically discoloured tissues in all parts of the palm.

Diseased roots are present on affected palms and E. Laville indicated (in correspondence, 1966) that four or five diseased roots among the 800 to 1 000 adventitious roots on a large palm are sufficient to ensure infection. Bulit et al (66) amplified the information on root infection. Malençon (265) and Péreau-Leroy (351) suggested that death might result from some toxic effect of the fungus rather than from simple congestion of the vascular tissues. They thought this might explain why the offshoots of a diseased palm may persist and develop for up to 10 years after the death of the parent.

Pathogen

The causal organism, F. oxysporum var. albedinis, occurs as a hyaline mycelium between and within the cells in infected, discoloured tissues, particularly in the base of the rachis. Microconidia usually occur in groups in vessels and other cells, but they may occur singly.

In culture (66, 351), this fungus produces a rose to violet stain in the medium. Great numbers of spherical to elongated and slightly curved, mostly unicellular hyaline microconidia are formed, measuring 3-15 x 3-5 μ. The falcate macroconidia are usually three-septate and measure 20-35 x 3-5 μ. The intercalary and terminal chlamydospores are spherical and occur singly or in groups of two or three. Sclerotia are dark blue to black, one to two mm in diameter and occur rarely.

The optimum temperature range for growth of the fungus is 21°C to 27.5°C but it grows reasonably well in temperatures as low as 18° or as high as 32°C. Thus, irrigated soils in areas such as Ksar el Souk, Morocco, maintain favourable temperatures for development of the fungus from May through October. The fungus may sporulate abundantly on date trash on moist ground. Substantial differences in pathogenicity among strains of the fungus and among isolates from various parts of the same palm were reported by Bulit et al (66). They studied isolation of the fungus from soil, where it is found commonly at depths of 30 cm or more and where it may persist indefinitely.

Infection

The fungus is soil-borne and can be spread by water, wind, diseased offshoots, which carry soil from diseased areas, or infected date tissues, especially infected wood or pieces of rachis. The henna bush (Lawsonia inermis L.) and lucerne (Medicago sativa L.), often grown among date palms, are symptomless carriers of the fungus (66,161). Infection takes place mainly through the roots but Péreau-Leroy (351) and Malençon (264,265) were unable to find a satisfactory technique for consistently infecting palm roots and producing symptoms. Laville (242), working with seedlings, showed that Fusarium does attack roots and can enter the plant through the root system and cause death. Bulit et al (66) used artificially infected seedlings in studies of strains of F. oxysporum var. albedinis. Malençon (263) demonstrated floral infection by this fungus, but this mode of infection is rare and establishment of bayoud in the palm as a result of floral infection was not demonstrated.

The spread of bayoud to isolated oases and larger date-growing areas was ascribed by Malençon (265) to movement of the fungus in articles made of palm tissues, such as pack-saddles, baskets and ropes, and Péreau-Leroy (351) found that the pathogen persisted for at least several weeks in infected rachis tissues. When such articles are discarded in a moist place favourable to the fungus, it can resume growth, become established in the soil and infect palms. Bayoud assumes epidemic proportions only when a combination of the pathogen, a susceptible variety, and abundant surface irrigation water exist.

Laville and Lossois (244) made a statistical study of the spread of bayoud at In Salah, Algeria, and discussed the factors involved. The progress of bayoud in a planting of several hectares of Bou Feggous at Zagora, Morocco, was studied by Bulit et al (66). From a small initial infection in 1956, the disease spread to 9 percent of the palms in 1958. By 1964, 60 percent (340 palms) were affected and most were dead. This planting had been cropped with lucerne and henna at the time of the initial infection; these crops require large amounts of irrigation water in summer, which would favour rapid spread of the fungus on susceptible crops.

Diseased palms of susceptible varieties probably never recover, but the advance of infections may be retarded by drastically reducing irrigation. This, however, renders the palm unfruitful so that the crop is lost either through drought or disease.

Control

The use of resistant varieties is the most practical means of controlling bayoud. Among six resistant varieties listed by Péreau-Leroy (351) only Takerboucht and Bou Ijjou were of acceptable quality although certainly not equal to Deglet Noor or Medjool. Breeding might combine resistance to bayoud with the desirable fruit and growth characters of some of the superior commercial varieties. A long-term programme for breeding and selection of superior, resistant date varieties for local use and export was begun in Morocco by Péreau-Leroy (350, 351) and, after a period of inactivity, has been renewed and augmented (66,161). Successful utilization of small offshoots would facilitate propagation of desirable resistant varieties; Bulit et al (66) have established small offshoots weighing two to 2.5 kg with up to 55 percent survival.

Meanwhile, moderate yields of dates and some mitigation of bayoud may be obtained by careful irrigation and the planting of resistant or semiresistant varieties. Irrigation of infested soil has a profound direct influence on the development of bayoud, and intercrops, especially henna and lucerne, may have an indirect influence by maintaining high levels of inoculum, according to E. Laville (correspondence, 1966) and Bulit et al (66). Palms of susceptible varieties growing in infested soil cannot withstand surface irrigation at any season but may be grown successfully on sites having a permanent water table at a depth of one to two m.

Semiresistant varieties may be grown successfully if surface irrigation is used primarily for intercrops and applied between October and May only. Resistant varieties tolerate surface irrigation and intercrops at all seasons.

The need for strict quarantine measures to prevent or delay the spread of bayoud and, at the same time, other pests and diseases of the date palm, has been recognized by Delassus and Pasquier (111), Maire (261), Crossa-Raynaud (101) and others. The first FAO International Technical Meeting on Date Production and Processing (157) recommended enforcement of quarantine measures against bayoud. Algeria (424) enacted quarantine legislation in 1942 and 1949, and strong quarantine measures have been enforced in the Sahara and North Africa (72). Mauritania adopted quarantine measures against bayoud in 1953 (294). The United States (431) and some other date-growing countries have quarantine regulations covering the movement of date palms and date palm products.

Khamedj

Introduction

Khamedj, or inflorescence rot, is present in date-growing areas in North Africa from Morocco to Libya (84,267). It has also been reported from Iraq (18,210,243), Italy (84A,168), Mauritania (293), Palestine (372) and Tunisia (101). Although inflorescence rots had been known for a long time, the first technical reports on this disease were published in 1925 by Cavara (82,83), followed by those of Chabrolin (84,84A). Khamedj is a general term for inflorescence rots (72) but refers primarily to that caused by the fungus Mauginiella scaettae Cav. (72,372). Inflorescence rots of minor importance are caused by species of Ceratocystis (see black scorch) and Fusarium (60).

Economic Importance

Khamedj is a common but usually minor disease problem; however, losses may occasionally be substantial. M. scaettae is destructive principally in areas with excessive or prolonged winter and spring rains, on neglected palms in marginal areas of oases, or in waterlogged soil, salty depressions or low-lying lands. Male palms, which frequently grow in these marginal areas and are communal property, may be particularly heavily infected because they receive little attention, except as pollen sources, the old inflorescences not being removed regularly. The fungus can survive in all parts of infected inflorescences and, if these are not destroyed, it may attack and ruin part or all of the new inflorescences in years favourable to the disease. Fortunately, the affected palms, both male and female, are usually scattered and the disease does not seem to spread rapidly.

Chabrolin (84) estimated that the losses might average 30 to 40 kg of fruit per year on severely affected palms. Trabut (434) ascertained that 5 percent of the palms were affected by khamedj in the southern Constantine area of Algeria. Chabrolin (84A) indicated that up to 10 percent of the palms might be affected in plantings in heavy, wet soil. The disease is so important in some Saharan oases (84) that the inhabitants developed empirical means of control, including sanitation and crude chemical treatments, prior to the establishment of modern control measures. Inflorescence rot is the only date disease of economic importance (18,243) among the 30 million palms in the Tigris and Euphrates valleys, principally between Baghdad and the Persian Gulf. In Iraq, important outbreaks are sporadic and occur only after prolonged cold, humid weather in the winter. Fawcett (165) considered inflorescence rot the most important disease of date palms in Tunisia.

Symptoms

The disease appears when the spathes begin to emerge in late winter or early spring. At first, affected spathes are barely distinguishable from normal ones. Brownish or rusty areas develop on the unopened spathe after the fungus has already invaded the floral tissues. Lesions may be confluent and are most common near the top of the spathe, which at the time of infection is soft and still hidden in the leaf bases. The internal face of the spathe under the lesions is yellow and translucent and may show brown dots corresponding to points of contact with diseased flowers. The fungus attacks flowers and strands and may move on to the stalk of the inflorescence. Spathes severely damaged when young may remain closed. However, the spathe usually splits and reveals partial to nearly complete involvement of the flowers and strands. Typically, certain infected palms develop symptoms every year, whereas others in the same planting show symptoms in the inflorescence only occasionally, even under conditions favourable for disease development.

Pathogen

In culture, *M. scaettae* develops a flat, white to cream-coloured mycelial mat that is soon covered with chains of hyaline conidia which give it a powdery appearance (84). The conidial chains fragment into units of one to several cells about 10-50 µ long and 5-10 µ wide. One- and two-celled units predominate. The same type of sporulation occurs on infected palm tissues. Germination is much better in tissue extracts or decoctions than in water.

On infected tissues, the fungus forms a distinctive white mycelial weft which later disappears, leaving a whitish bloom of spores on the surface. Spores are considered to be short-lived and not of much importance in the persistence of the fungus, which apparently survives in old tissues as mycelium. The fungus, which is capable of direct penetration of the spathes, is an agressive parasite. The role of contaminated pollen in the spread of the disease seems to have been ignored and as R/ayner ?7(372) pointed out, no detailed studies of dissemination, infection and host-parasite relationships have been reported. He considered that *M. scaettae* would not spread readily into uncontaminated areas because the spores are short-lived and transportation of infected tissues would be negligible. Nevertheless, in France, Anselme and Baltzakis (20) found this pathogen occurring as a rot of packed date fruits where a piece of infected strand was present. Fancy packs of Deglet Noor dates from North Africa often contain a piece of strand for decoration.

Control

Good sanitation is the first step in the control of khamedj. All diseased parts of the inflorescence should be collected and burned immediately after harvest, and in all palms the old flower stalks and spathes should be cut out as completely as possible and removed from the planting. Fungicides should be applied immediately after harvest on the old flowering area of diseased palms, and in early spring on the new flowering area just before the spathes emerge. Chabrolin (84) reported good control by spraying with bordeaux mixture or a strong lime-sulphur solution, or by dusting with a copper sulphate-lime mixture, essentially powdered bordeaux mixture. Laville (243) recommended sprays of dichlone and thiram[1/] at 0.4 and 0.3 percent, respectively, at a rate of about 8 l per palm. The dust was cheaper and more easily applied than the spray in small plantings. According to Teisseire (424), the Ghars variety is more susceptible than Deglet Noor, and Tafezouine, Hamrain and Takermest are resistant. Zahidi and Hillawi are somewhat resistant, but Khadrawy and Sayer may be severely attacked (243).

Fruit Rots

Distribution and Economic Importance

Preharvest fruit rots may be caused by one or more of a large number of fungi (59, 166), and occur wherever dates are grown. The economic importance of fruit rots varies greatly, since their incidence is governed by the occurrence of rain and high humidity during the khalal and later stages of ripening. Darley and Wilbur (106) reviewed fruit spoilage in California. Although losses of 25 percent or more occurred sporadically in individual gardens or districts, they estimated the annual loss from fungus spoilage at about 5 percent when paper covers, fungicidal dusts and spreader rings were used. When the cost of controlling fruit rots is added to the loss of fruit, they become the most important disease problem of dates in California.

Estimates of loss from fruit rots are seldom reported from the date-growing areas of the Old World, although such rots do occur. Kearney (226) noted that rains at harvest time in Tunisia sometimes caused serious losses of fruit. To circumvent rain damage, dates are often harvested either individually or in bunches while in the late khalal to the early rutab stages, and the fruits are then ripened on mats, on racks and in other ways (72, 130, 129, 277). In Algeria (72) late maturing varieties like Deglet Noor may be damaged by rain and fruit rots in unfavourable years with losses reaching 25 to 30 percent. In Israel (90) fruit rots occur in late varieties such as Deglet Noor and Barhee.

Pathogens and Disease Development

Species of _Alternaria_, _Helminthosporium_ and _Macrosporium_ may infect fruits directly, beginning in the khalal stage. Fawcett and Klotz (166) also found _Citromyces ramosus_ Bain and Sart. and _Aspergillus niger_ van Tiegh. causing a

---

1/ Martin, H. 1957. Guide to the chemicals used in crop protection. 3rd ed. Canada Dep. Agric.: London, Ontario, 306 p.

calyx-end rot in the late khalal or early tamar stages. Rieuf (388) discussed A. niger and A. phoenicis (Cda.) Thom in relation to date fruit rots. These and numerous other species of saprophytic fungi, among which yeasts are important, also invade ripening fruits through wounds, especially during periods of rain and high humidity when fruits may split and crack and provide easy entry for fungi. The Deglet Noor variety is especially liable to such injury. The discussion by Nixon (333) of the factors involved in fruit rots is of general application in date culture.

Control

Losses from fruit rots may be reduced by several means (106,333). In the United States, the following procedures, used alone or in combination and mostly on Deglet Noor palms, have been effective in reducing fruit rots. Prior to the khalal stage, wire rings may be inserted in the bunches to promote good ventilation and drying of wet fruit. Even in the absence of rain, dew may form on low-hanging fruits of young palms and rather humid conditions may prevail because of weeds or cover crops in association with soil moisture. During the early khalal stage, a dust consisting of 5 percent ferbam and 95 percent dusting sulphur, available under several trade names (333), may be applied to the bunches to reduce fungal spoilage; the dust also discourages some insects (41) At the same stage of growth the bunches of Deglet Noor are covered with paper wraps or bags to exclude rain. Fruit rots may be aggravated by cultural conditions that increase humidity, such as weeds, intercrops and standing water, and these conditions should be taken into account in planting and maintaining date orchards.

Graphiola Leaf Spot

Distribution and Economic Importance

Graphiola leaf spot is caused by Graphiola phoenicis (Moug.) Poit. It is the most widespread disease of date palms and probably occurs wherever the palm is cultivated under humid conditions (166) In most date-growing areas where Graphiola is common, the palms are frequently seedlings or of inferior varieties (330), and yields may be reduced by premature death of the leaves. Dates are, nevertheless, an important crop even under such conditions.

Graphiola leaf spot is common on palms in the Punjab (277) where rainfall ranges from 150 to 600 mm during March to August, inclusive. Fawcett (165) found leaf spot to be the most common disease of dates in the Arab Republic of Egypt, especially in the Delta and at Fayoum. He believed that the commonly severe infections near the coast "must interfere considerably with the growth and production of the palms". In the Arab Republic of Egypt old and diseased leaves are removed as a sanitary measure; Fawcett (165) however, implied that excessive pruning could reduce date yields. Leaf spot often occurs in the humid coastal areas of North Africa and the Near East(267,329,330,332) but the disease is of little importance in the principal interior date-growing areas of North Africa (84A). Graphiola causes serious damage to palms in several of the southernmost, humid date-growing areas of Senegal, Mauritania, Mali and Niger (266).

Pathogen and Disease Development

G. phoenicis was described by Fischer (168) and Killian (232) established its taxonomic position as one of the smut fungi. The fungus develops subepidermally in small spots on both sides of the pinnae and on the rachis. Finally, the fruiting structures emerge as little black, covered sori or pustules and are often numerous.

The sori are one to three mm in diameter and consist of two layers; the outer peridium is hard, dark and persistent, but the inner hyaline peridium is thin and degenerates after the spores mature. Spores are produced in fertile areas of the sori and are interspersed with groups of sterile filaments. At maturity, the sori open to liberate masses of yellow spores. Individual spores are spherical to ellipsoidal, 3-6 μ in diameter, and have a thick, smooth, hyaline wall. After the spores are disseminated, only the rough black craters of the sori remain. Heavily infected leaves die prematurely.

Control

Leaf pruning (165,277) is the usual practice for reducing damage from Graphiola leaf spot. Spraying with bordeaux mixture has been recommended by Milne (270) and later authors. Nixon (330) studied the relative susceptibility of date varieties in Texas and concluded that a few varieties, especially Kustawy, were tolerant to Graphiola. Tolerant varieties might be useful either directly or in breeding improved varieties.

Black Scorch

Black scorch, which includes medjnoon or fool disease (204) is a minor disease of sporadic occurrence. It is caused by Ceratocystis paradoxa (Dade) C. Moreau. The fungus is normally present on date palms only in the imperfect form (Thielaviopsis). Black scorch has been reported from Tunisia and Algeria (165), the Arab Republic of Egypt (62,165), Saudi Arabia (329), Iraq(243) and the United States (76). In Mauritania, Brun and Laville (64) found this fungus associated with the bending head disease described by Munier (294). No estimates of loss are available.

The investigations of Klotz and Fawcett 234 and Streets (415) provide the principal information on black scorch. All parts of the palm are susceptible to infection by C. paradoxa, but symptoms are usually expressed in four fairly distinct forms: as a black scorch on the leaves, as an inflorescence blight, as a heart or trunk rot, as a bud rot on palms of all ages. The infections are all characterized by slight to extensive necrosis of the tissues. Bud rot and heart rot, though uncommon, are the most damaging manifestations of C. paradoxa and may kill the plant. Some palms recover, probably by development of a lateral bud, and these show a characteristic bend in the region of infection. This is the medjnoon or fool disease symptom.

In Iraq this fungus is associated with a decline and death of palms in the Shatt el Arab region, where Laville (243) believes that infection is associated with sudden changes in the level and salinity of the water.

The imperfect form of C. paradoxa develops rapidly on host tissues, or in culture, and produces long chains of endospores which fragment readily. Two types of spores are produced: hyaline, cylindrical microconidia, measuring 5-15 x 3-7 μ, and dark, egg-shaped macroconidia, measuring 11-17 x 7-15 μ. Between the two types may be found all gradations in size and shape and in intensity of brown colouration.

Infected tissues become blackish brown, then shrivel and form conspicuous lesions with a scorched, carbonaceous appearance. Dark conidia may develop abundantly on the lesions. With bud rot and inflorescence rot, where the tissues invaded are relatively succulent, secondary organisms may develop rapidly and augment the damage. Black scorch has been observed on 17 date varieties. No infections were found on the Tazizoot variety, and Deglet Noor seemed less affected than other commercial varieties; Thoory was the most susceptible. C. paradoxa may be partly controlled by removing and burning all infected leaves and inflorescences. Dusting of the area where the inflorescences emerge with powdered bordeaux mixture in early spring is a suggested control measure.

Diplodia Disease

Diplodia disease is a minor disorder caused by *Diplodia phoenicum* (Sacc.) Fawc. and Klotz(166) in the United States, although a second species, *D. natalensis* P. Evans, was reported on date palms in Israel(278). The disease has been reported also from the Arab Republic of Egypt and Tunisia (165). No estimate of loss from *Diplodia* disease are available. Fawcett (166)investigated this disease in California, where at one time it caused considerable damage. The fungus may infect and even kill young offshoots at the base of palms; it may infect the outside leaves and then finally kill the younger leaves and terminal bud, or the central leaf cluster may die before the older leaves. Yellowish brown streaks 15 cm to about one m in length extend along the leaf base and rachis. The lesions become brownish; internal infection and necrosis may be extensive in leaf tissues. The disease was observed on 17 varieties in California.

The most common spores of *D. phoenicum* in cultures or in dead tissues are hyaline. Mingled with them are the typical dark, septate spores of *Diplodia* which measure 22-24 x 10-12 μ. The light-coloured spores are somewhat larger and mostly unicellular. Dark, intercalary chlamydospores are produced abundantly in cultures. Pycnidia may develop on dead leaf tissues.

Fawcett(166) believed that the fungus usually enters palms through natural or artificial wounds. Faulty irrigation, causing some roots to die back to the base of the palm, may be associated with the disease. Recommended control measures consist of dipping diseased offshoots, or those suspected of being infected, in ammoniacal copper carbonate solution and annual spraying of offshoots with the same fungicide. Disinfection of pruning tools is suggested.

Omphalia Root Rot

Omphalia root rot(40,229,230) is a minor disease of date palms in the United States. The causal fungi, *Omphalia tralucida* Bliss and *O. pigmentata* Bliss, are fairly widely distributed in palm plantings in the Coachella Valley of California and have been isolated from the roots of many varieties and seedlings. Although *Omphalia* spp. are pathogenic on palms (40) apparently no constant association exists between the presence of these fungi and visible decline symptoms. The fungi have been isolated from both healthy and declining palms, but not all declining palms - as indicated by loss of vigour, stunting and unfruitfulness - yielded these fungi (229). Poor cultural practices apparently produce identical decline symptoms, especially in Deglet Noor, which at one time was considered especially susceptible to Omphalia. Kenknight's (229) data showed plantings of Deglet Noor to have a lower percentage of infected palms than those of any other commercial variety at the time of his survey, 1945-1948. In 1965, Sachs (397) found both species of Omphalia in Mauritania on four palm varieties. He reported studies on the fungi and their control.

Belaat

Belaat refers to a minor disease of the date palm in Algeria caused by Phytophthora sp. It was first described by Maire and Malençon (262) and Maire (261) published a more detailed account. Monciero (283) and Calcat (72) have discussed it more recently. No estimates of the losses caused by belaat have been published.

Belaat usually occurs in neglected plantings. On affected palms the entire central cluster of young leaves turns white fairly rapidly. Infection occurs near the growing point in the form of a wet heart rot that kills the bud and the bases of the youngest leaves. The infection may be arrested a short distance below the bud or continue downward. Secondary organisms rapidly follow the initial infection. Some affected palms recover by producing a lateral bud. Offshoots of diseased palms usually remain healthy.

Blacknose

Blacknose is a physiological disorder of date fruits reported from the United States and the Arab Republic of Egypt. The melanose described by Munier (294) appears to be the same disorder and is reported to occur in Mauritania and throughout North Africa. Although no official statistics are available, in California the annual loss from blacknose on Deglet Noor dates may approach 5 percent of the crop. About 80 percent of the acreage in California is planted to this variety. Exceptional losses of 50 percent or more in individual plantings have been reported (304). In the Arab Republic of Egypt (62) Hayani is the most susceptible variety and the one most widely grown.

Blacknose results from excessive checking of the epidermis, especially in the form of transverse cracks at the stylar end of the fruit, and subsequent drying and deterioration of the underlying tissues. Aldrich et al (12) summarized the literature on blacknose and reported on physiological investigations of checking in Deglet Noor dates. Nixon (333) has given a more recent account. Checking is induced by high humidity and rainfall at the late green stage just before the fruit acquires its khalal colour. Therefore, conditions tending to increase humidity, such as standing water, excessive soil moisture, presence of intercrops and weeds should be avoided, especially at this stage of fruit development. Overthinning can also increase the incidence of checking and subsequent development of blacknose. Careful attention to irrigation, cultivation, thinning and ventilation may mitigate the damage.

Le Coeur Qui Penche

"Le Coeur qui penche", or bending head, is an uncommon but fatal disease of unknown cause reported from Tunisia (204) the Arab Republic of Egypt ([illegible]) and Mauritania (294), where Brun and Laville (64) found *Thielaviopsis paradoxa* and *Botryodiplodia theobromae* Pat. associated with it. They suggested that these fungi are essentially secondary parasites, attacking palms weakened by drought or poor cultural practices.

The symptom from which the common name derives (294) takes the form of a grouping of the terminal leaves into an erect fascicle with a bent tip. These leaves eventually die and fall over. Meanwhile, older leaves die and hang vertically. These symptoms are commonly associated with a basal heart rot in the palm. Where a bud rot or infection of the upper stem occurs, the leaves may die and fall more or less to one side. The symptoms may be intermediate and are not always diagnostic of the type of rot involved. In any case, after the palm dies the trunk bends and may break. Munier (294) recommended that, to limit spread of the disease, all portions of infected palms be burned on the site.

Barhee Disorder

Barhee disorder is the name given to a bending of the crown of Barhee date palms (105,108) that occurs in most of the fruiting plantings of this variety in California. Affected palms are usually over 3 m high and bend over to an angle of about 5° to nearly 90° from the perpendicular; 80 percent of them lean to the south. About 50 percent of the palms in a 10 to 33-year old planting have been affected for at least six years. The number of fruiting bunches is reduced in direct relation to severity of bending. Affected palms may recover and resume upright growth. Neither the cause nor the control of this disorder is known. Barhee disorder does not appear to be related to "le coeur qui penche" (64, 123, 294). The Barhee palms in California are all derived from a single importation from Basrah, Iraq, in 1913.

Crosscuts

Crosscuts, which appear as clean breaks in the tissues of the lower part of fruit stalks on palms, may vary from a slight notch to complete breakage of the stalk. Fruits on strands in line with the break wither and fail to mature properly. Sayer is especially susceptible to this disorder, and in the United States up to one fourth of the crop may be lost. In 1934, about 1 000 fruit bunches were lost because of crosscuts in a single planting in California. Khadrawy is also susceptible to crosscuts. Bliss (39)concluded that crosscuts resulted from an anatomical defect in the fruit stalk involving internal, sterile cavities or fractures that led to mechanical breakage during elongation of the stalk. Crosscuts are most commonly found in varieties having crowded leaf bases and their incidence increases as the palms age. Crosscuts may be avoided by using non-susceptible varieties, and losses may be reduced by leaving a few extra fruit stalks per palm in susceptible varieties(39). Similar crosscuts or v-cuts occur occasionally in leaves (166).

Faroun

This is a local name for a rapid and fatal date palm decline of unknown cause reported from Mauritania in 1967 by Laville and Sachs (245) Female and male palms are affected. The first symptom is a failure of apparently normal palms to flower for one or two seasons before foliage symptoms appear. Affected trees become conspicuous when they assume a parasol form produced by the old and middle-rank leaves being held somewhat horizontally. Subsequent new leaves and their pinnae are reduced and distorted, until finally the central portion of the crown becomes a grotesquely stunted rosette. These symptoms are accompanied by abortion of the axillary buds, which would account for failure of the palms to flower. Offshoots are affected on some diseased palms. Internally, palms in advanced stages of decline have numerous brown gum pockets and long dark-coloured cracks in the crown tissues. Similar cracks occur in leaf bases near the point of attachment to the trunk. No causal agent has been identified; an insect, *Piezodorus pallescens* (no authority in text), has been found fairly constantly in the crown tissues of affected palms.

Rapid Decline

Rapid decline, or rhizosis, is a minor but fatal disease of unknown cause which has been observed on date palms in California for more than 30 years. Darley and Wilbur (107) reviewed previous work on this disease and reported new investigations. The onset of symptoms is rapid. Typically, vigorous female palms in full production suddenly shed a large part of the immature fruits in summer but, if the palm is affected later, the fruits shrivel on the bunch. A reddish brown discolouration occurs on pinnae of the oldest leaves and almost simultaneously the unexpanded young leaves wilt. Mature leaves die rapidly and progressively toward the top of the palm. Offshoots, if any, usually die with the parent plant and thus the disease tends to be self-limiting. Male palms also are affected. There is no indication of a disease pattern, since single affected palms occur at random. No varietal resistance has been noted. Serious losses have occurred in only a few plantings.

Dry Bone

Dry bone is a minor disease of the date palm reported from the Arab Republic of Egypt and Tunisia (165) and from the United States (165) Fawcett and Klotz (166) thought it might be caused by a bacterium. The principal symptoms on the leafstalks, midribs and pinnae are whitish, irregular blotches and streaks that become sharply delimited and may have reddish brown margins. The lesions vary from one to several cm in greatest diameter, but involve only the epidermis and a thin layer of subjacent tissue.

Black Scald

According to Nixon (333), black scald is a minor date fruit disease of unknown cause occurring in the United States. Well-defined blackened areas appear on the tip and sides of the fruit and affected tissues have a somewhat bitter taste. Black scald does not seem to be related to blacknose.

Al-Wijam

Al-Wijam is a minor date palm disease of unknown cause occurring in Saudi Arabia, according to Nixon (329). Affected palms are stunted, become unfruitful and eventually die. The only unique symptom is "a faint, narrow, yellow, longitudinal line" on the midribs of the younger leaves. The decline occurs whether conditions for growth and fruiting are favourable or unfavourable.

CHAPTER XII

THE IMPROVEMENT OF PRODUCTIVITY

In the preceding chapters the various aspects of date production have been discussed. In conclusion, a review is made of the practical steps that may be undertaken to improve productivity in North African and Near Eastern countries.

A. YIELD

The yield of most bearing palms in the region is something between five and 50 kg per palm per annum. This was confirmed in 1975 by delegates to the Third FAO Technical Conference on the Improvement of Date Production, Processing and Marketing. The statistics they reported indicated national average yields for example of about 17 kg for Iraq, 15 kg for Iran, 20 kg for Libya, 15 - 20 kg for Morocco. Previous reports from these and other countries of the region confirm that this has been the general level of production in recent years.

Well cared for palms produce much higher yields of course. FAO experts (34A) reported in 1974 that date palms at Jahrum, Iran (at elevations around 1 000 m) often produced 100 kg each. Dowson (126) in 1961 reported that yield of well cared for palms in Libya was probably not less than 50 kg. Experiments at Zaafaraniya Station, Iraq, reported at the Second FAO Date Conference (7) that average yields over four years was over 100 kg per palm.

Since the planting distance is so variable, it would be more scientific to give yield in terms of kg per ha, but statistics in these terms are not always available. Ali Hussain at the Second FAO Date Conference in 1965 reported that the average yield per donum in Iraq ranged between 580 and 1 072 kg which would give a mean of 757.17 kg/donum or 3 028.68 kg per ha, say, three tons per ha.

Weighing the yield of each palm is desirable so that a permanent and continuous record can be maintained of each palm's history. Basra, Iraq.

The best examples of sustained high yield, both per palm and per hectare come from California. Hedlund (199) reported that from 1960 to 1964 yields exceeded five tons per acre ($12\frac{1}{2}$ t/ha) which at 50 palms per acre represents an average of 100 kg per palm. Subsequent reports indicate that this average yield has since been maintained.

Over 47 years ago the records from a private farm, the Jarvis Ranch (218) exemplify the good results obtained from palms of different ages, summarized in the Table below.

Table 18

YIELD IN 1932 OF THE DEGLET NOOR CROP,

JARVIS RANCH, COACHELLA VALLEY, CALIFORNIA

Average Age of Palms in Years	Yield in kg
4	11
5	41
6	52
7	91
8	122
10	170

It is clear that the production techniques evolved in California have been proved capable of giving high and regular yields under the prevailing conditions there. The problem for the Old World is how these techniques may be adapted to suit local climatic and economic conditions. To some extent this is a task for research, but there is also a considerable wealth of local experience that may be put into practice. The causes of low yield are in fact well known, although it is not always easy to correct them!

The possibilities for improvement will vary according to the size and character of the holding. The first distinction that must be made is between the independent small holding and the large plantation - whether or not the large plantation is under one ownership or run under some form of cooperative management with numerous individual tenants.

B. SMALL PLANTATIONS

The independent small holder will not be able to change his traditional cultural methods to any great extent and his palms must be climbed several times a year with the aid of a climbing belt to prune the fronds, pollinate the spathes as they open and harvest the bunches as they ripen. If he is a comparatively prosperous small holder he will be interested in obtaining a mechanical pump to raise his irrigation water and perhaps a rotary cultivator or two-wheeled tractor to prepare his vegetable plots, but he will probably continue to rely on a lot of hand work with his hoe to manage his irrigation and keep his plots free from weeds. He can be helped in financial ways of course - by loans to buy pump and motor-cultivator and by subsidized or price-controlled fertilizer supplies - but, on the technical side, what he would probably appreciate most is advice and assistance in controlling pests, especially the mite (*Olygonychus*), the moth (*Batrachedra amydraula*), the dubas bug (*Ommatissus*) and in some regions also the scale (*Parlatoria blanchardi*).

The small holder will in any case be very dependent on his inter-crops for a regular income and may often regard his dates as of secondary importance. Some attention may be necessary to thinning out so as to devote the limited space to whatever pays best - whether citrus, vegetables or fodder crops - and to arrange spacing so that all crops get at least the minimum of light to grow well.

Whatever the size of the holding, improvements in lay-out must be accompanied by adequate irrigation and manuring and where inter-crops are grown, these will determine the frequency and rates of application.

C. LARGE PLANTATIONS

1. Preliminary Conditions

For large plantations, considerations are rather different. Before any other decision is made, it must be ascertained, by analysis of present and probable future costs and returns, if the palms are worth keeping. If they are not, the chain-saw and bulldozer are the best answer, and this is often the case on valuable irrigable land with irregularly planted palms. It may also be found that the palms have grown too tall to make it economic, or even safe, to climb them several times a year and quite impossible to reach the heads by ladder. Again, the palms may be seedlings or of a poor variety and/or include too many males. Or yet, again, the local climate may not be really suitable for dates. It is obviously no use spending time and money on improvements if, for example, rain or humidity regularly spoils the crop.

One consideration that may make it impossible to improve a plantation is if the whole area is badly drained and no improvement in drainage on the individual holding is possible. Conversely, if good drainage outlets are available, improvements to the drainage system will be among the first considerations in a programme of improvement 1/.

It will usually be necessary to consider the question of irrigation improvement at the same time, with due regard to leaching requirements as discussed in Chapter VIII. If the holding is to be mechanized, some form of surface irrigation will be required in any case.

Proper calculation of the amount of water required and adequate control of the rates, timing and mode of application are fundamental to good management.

1/ In the UNESCO/UNDP Project in Tunisia (TUN/5), previously referred to, remarkable results were obtained by improvements to irrigation and drainage of a plantation of mature Deglet Noor palms. In 1964, yields on the four plots concerned ranged from 2.93 to 3.47 t/ha, with an irrigation equivalent to 3.5 to 4 mm/day. By 1968 yields had progressively increased to a corresponding range of 8.55 to 10.47 t/ha with irrigations equivalent to 6.75 mm and 10 mm/day. (Technical Report UNESCO/UNDP 55, TUN/5, Paris, 1970).

2. Spacing

Where palms are regularly planted and reasonably healthy but merely overcrowded, a process of thinning-out may be considered. In this case it may prove convenient to leave double rows and form a series of "bays" by uprooting two or three rows between each double row retained, thus creating convenient areas for growing vegetables and fodder crops. The value of palms as a wind-break on the side of the holding nearest prevailing wind should not be overlooked.

3. Mechanization

In all large plantations, the need for at least partial mechanization must be considered for the following operations:

(a) Transport - It is uneconomic to carry the crop, date fronds, manure, fertilizers, etc. by hand or by donkey - a tractor and trailer is, therefore, required and the plantation must be sufficiently open to allow free passage. In very large plantations, roads will be needed to allow the entry of lorries.

(b) Cultivation and Weed Control - Land surfaces must be levelled and furrows drawn for irrigation and at least light cultivation sill be needed for suppressing weeds and harrowing-in fertilizers. For all these jobs, a four-wheeled tractor and tractor-drawn implements are required. Weed suppression is not particularly important for small annual weeds but Halfa (*Imperata cylindrica* (L.) P. Beauv.) is a terrible enemy and must be eradicated "at all costs" - but the operation will probably not be economic without a powerful tractor.

(c) Pollination, Pruning and Harvesting - Economic production can no longer be based on the climbing belt in most countries. If capital is not available for mechanical booms, light-weight metal ladders will be required. It will, therefore, be inevitable that very tall palms will become uneconomic - indeed this problem may have to be considered any time from the 30th year on, even although the palms have, by then, barely reached "middle age". Once this problem has been overcome, it will be quicker and simpler to use a pollen-blower than hand pollination methods and losses of crop through inadequate or infrequent pollination will be easier to avoid.

Improvement in harvesting methods is likely to yield the most immediate financial advantages and in any case it is the essential first step for the elimination of pest infestation of the crop and for the improvements in packing and processing required for a high-quality product.

4. Propagation and Rotation

Because of the tendency of palms to become uneconomic for mechanized production as they grow older, a modern plantation should plan regular small annual plantings or a larger planting every five years, so as to have blocks of successive ages and a constant supply of new planting material. Where large areas of plantations have become uneconomic due to the age of the palms, energetic measures will often be required to bring in large numbers of offshoots from younger "mother" palms in another district. The new plantations should be laid out in a way suitable for mechanization - i.e. with suitable varieties, surface irrigation, good drainage, a minimum of open drainage canals, wide spacing and adequate roads. As the capital investment will be a high one and as returns from the palms will be minimal for the first four to seven years, careful consideration must be given to the choice of inter-crop during this period. Financial, as well as practical, considerations will make it desirable to establish the date plantation progressively in five-year blocks. This will overcome much of the difficulty of finding good new offshoots as well as providing for progressive rejuvenation of the plantation in later years.

APPENDIX I

JANI AN-NAHLAH FI KAYFIYYAT GHARS AN-NAHHLAH

(THE HARVEST OF THE BEE IN DATE CULTURE)

by

Amin bin Hasan Hulwani Al-Madani,

Teacher in the Haram an-Nabawi (the Prophet's Mosque)

May God Forgive Him !

Hasaniyah Press, Medina, 1886 (A.D.)

(Translated into English by V.H.W. Dowson and Abbud Ali esh-Shubbar)

In the Name of God, the Merciful, the Beneficent

The purpose of this article on planting date palms is that they may be of all the palms in the world the strongest and stoutest and heaviest yielding and so firmly fixed in the earth that no sotrm can uproot them. Know then that date planting in the olden times was known only to the inhabitants of al-Madinah, and that date palms were their dearest possessions, and that it was of them that they used to boast; but, with the passing of the years, they forgot how to plant correctly; and they began to plant anyhow in the style of Khaybar and Egypt and Morocco and other places, where care is not taken of palms. Then, in the year 1260 A.H. (1844-5-A.D.) the people of al-Madinah woke up to this fact, and learnt again how planting should be carried out from the people of al-Qasim, that is to say, from the people of Buraydah and "Unayzah and ar-Rass".

There are seven rules to be observed in the correct cultivation of date palms.

The First Rule

Dig a hole one metre cube. This hole is called by the Hijazis faqrah (hole). Return two-thirds of the e arth to the hole and leave one-third unreturned. Then plant the shoot. (Note. V.H.W. Dowson. The word used is Sanu, which is not given as "shoot" in Fayruzabadi, Salmoné, Hava, or Wortabet, though the first gives Sanwah as "two date palms", so the translation employed above is a guess.)

Then water the shoot with a little water, so that the heart is not drowned, daily for sixty days, until it is apparent that the shoot has struck and put out new roots in the earth surrounding it, and produced small new fronds, whereupon another ten centimetres of earth should be filled into the hole. Do likewise, as new fronds appear and the shoot grows, until you are certain that the shoot has become strong, and struck, and produced many fonds. At this stage, watering may be heavy, because there is no further fear that the shoot may be drowned.

The Second Rule

Space palms at least ten metres apart. We have found that closely planted palms do not yield heavily. Supposing an acre (of land) were to be planted with a hundred palms and another acre with fifty, it is certain that the yield of the fifty would be greater than that of the hundred.

Know that no crop stands so much and such continuous watering as dates, which are not hurt by daily watering, and whose vigour and yield, unlike those of other trees, increase as the water increases, be it sweet or salt. Much continuous, and salt, water turns (the leaves of) other trees yellow.

If you have planted (shoots) in the manner I have indicated, your palms will be stout, lusty, all their parts in proportion, and good bearers, such that ten of them will equal a hundred palms planted as are other palms in the world, and such that they will live long, and not be uprooted by violent wind, ir, of the wind is very violent, will rather break than be uprooted.

Heavy yield results from a large number of roots. This is confirmed by previous agricultural experts, who said that every increase of a root results in the increase of a pound of dates. The presence of a large number of roots depends on the presence of friable soil. (This is the reason for) first digging out a hole and then filling it up again (when planting shoots). There is no doubt that roots can penetrate better and more easily in loose earth than in hard ground. This we have seen with our own eyes in an experiment with palms newly planted by (the tomb of) our Lord Hanzah in al-Madinah the Illumined, and with older palms at Qaba; and there is no better guide than experiment.

The Third Rule

Choose a shoot for removal from a palm of known variety, which bears heavily of good dates, (such as, for example, the Hilwah, the Baydh, the Barni, and the Jaawi) and of which the 'ajwah (dates pitted and pressed into a block) has the taste of sweetmeat, because the date palm is like the sons of Adam who resemble their parents.

It is best to bring (shoots) from outside (al-Madinah), from for example, the mountain of Aja, Salma, Qasim or Yanbu', because the foreign (shoot), when brought from its home and properly looked after, succeeds and flourishes.

The Fourth Rule

The first year in which a palm bears, pollinate it so that sada (green dates) are produced, called in the dialect of Egypt nini. Then cutt off all the bunches and throw them away, because this removal of the first yield strengthens the heart of the palm. It will be found that, supposing a palm bears two bunches in its first year of yielding, and these are removed, the next year it will bear five in their place.

The Fifth Rule

This is very important. Do not allow shoots to remain at the base of the palm; but, on the contrary, as the palm produces shoots, kill them, because each one which is killed results in greater strength to the mother, for there is no doubt that the palm is more profitable than its young shoots, the value of which is unknown.

The Sixth Rule

Do not cut the fronds until they hang downwards of their own accord, and have begun to dry naturally. It is to be observed that the exceedingly severe treatment of the palms in the villages of Egypt owing to the sale of the fronds to the people of the graveyards is a cause of the weakness of the palms there. The profit on the sale of frond does not compensate for the weakening of the palm. Try and see.

The 'Arabs have a practice called pruning (tashdhiib), which consists in cutting away the lowest ring of fronds, when these have become numerous, and the lowest ring of fibre, when it has grown very thick; but, if too much is cut, the palm is hurt, just as a man would be if, in shaving his head, the skin as well as thei hair were removed. What would be his state? How many would be his pains! And how altered would be his condition! The date palm in many ways resembles man in his nature and characteristics, as shown in the book of Ibn Washshiyyah on the Science of Agriculture.

Verily the Prophet (on whom be the blessing of God and whom may God gree!) said, "Honor your paternal aunt, the date palm." He called it aunt because the paternal uncle and paternal aunt are a man's nearest relatives.

As for the manuring of the date palm, this is called Sibaakh (manuring or salting) in the Egyptian dialect, and, in the Hijazi dialect, tadmiin (dunging). It is the best thing for increasing the yield of the date palm, and for the production of date syrup. Usually, cow dung or camel droppings are used, and are buried beneath the palm in winter; but the best of all mabures is fermented dates, if spread beneath the palm in winter, and, indeed, it is the most potent elixir for the production of dates in the date palm. The easiest way to prepare the fermented dates is to bury jars during the summer, and to put into them the refuse and trodden-on and bird-pecked dates, and then to water them, and to close the jars, and let the mixture ferment, and remain till the winter, and then to apply three pounds of it under each palm; and, in this way no cost will be incurred, because, in any case, the reject dates are of no other use.

(The Seventh, and final, rule is concerned with irrigation.)

APPENDIX II

LIST OF SERVICES, INSTITUTIONS AND RESEARCH STATIONS INTERESTED IN THE DATE PALM

All of the following institutions have been interested in date production and/or protection at one time or another, but some may no longer be concerned with the crops. FAO would appreciate notification of any corrections to the list.

ALGERIA

- Station expérimentale agricole de Sidi-Mahdi
  B.P. 17, Touggourt

- Ministère de l'Agriculture et de la Réforme Agraire
  12 blvd. Colonel Amirouche, Alger

ARGENTINA

- Instituto Nacional de Tecnología Agropecuaria (INTA)
  Rivadavia 1430, Buenos Aires

- Instituto Nacional de Tecnología Agropecuaria (INTA)
  Estación Experimental agrícola
  Casilla de Correo 1, La Banda, Santiago del Estero

- Date Experimental Garden
  Cruz del Eje (Córdoba)

- Estación Experimental Regional Agropecuaria Famailla (INTA)
  Casilla de Correo 9, San Miguel de Tucuman

AUSTRALIA

- Agricultural Branch, Northern Territory Administration
  Darwin, Northern Territory

- Department of Agriculture, Division of Plant Industry
  Horticultural Branch, William Street, Brisbane 4000, Queensland

BRAZIL

- Instituto de Pesquisa Agropecuária do Leste (IPEAL)
  Caíxo Postal 552, Salvador, Bahia

CHAD

- Ministère de l'Agriculture et de l'Elevage
  Direction de l'Agriculture, B.P. 441, N'djamena

COLOMBIA

- Instituto Colombiano Agropecuario (ICA)
  (Programa nacional de Horticultura)
  Apartado Aéreo 7984, Bogotá

EGYPT (ARAB REPUBLIC OF)

- Ministry of Agriculture, Horticulture Division
  Gizeh, Cairo

- Alexandria University, Faculty of Agriculture
  Alexandria

- National Agricultural Research Centre
  Cairo

ETHIOPIA

- Institute of Agricultural Research
  P.O. Box 2003, Addis Ababa

- Alemaya Agricultural Experiment Station
  College of Agriculture HSIU
  P.O. Box 13, Dire Dawa

FRANCE

- Institut français de Recherches fruitières (IFAC)
  6 rue du Général Clergerie, Paris 16e

- Station centrale de Pathologie végétale (INRA)
  Route de St. Cyr, 78 Versailles (Yvelines)

INDIA

- Punjab Agricultural University
  Ludhiana, Punjab State

- Regional Fruit Research Station
  Abohar, Punjab State

- Fruit Research Station
  Khedoi, Gujarat State

- Fruit Research Station
  Shriganganagar, Udaipur, Rajasthan

IRAN

- Plant Pests and Diseases Research Institute
  P.O. Box 3178, Teheran

- Seed and Plant Improvement Institute
  Ministry of Agriculture and Natural Resources
  Teheran

- Department of Agriculture
  Ahwaz

- College of Agriculture
  Ahwaz

IRAQ

- Palm and Date Research Centre
  P.O. Box 2416, Baghdad

- Horticulture Division
  Horticulture Experiment Station
  Zaafaraniya, Baghdad

- Department of Entomology and Plant Pathology
  Baghdad University, Abu-Ghraib

- Iraqi Date Organization
  Baghdad

ISRAEL

- The Volcani Institute of Agricultural Research
  P.O. Box 6, Bet Dagan

- Department of Food and Biotechnology
  Technion, IIT, Haifa

- The Jordan Valley Research Committee
  Beth Yerach

- Ministry of Agriculture, Soil and Irrigation Extension Service
  Beth Shean

LIBYA

- Agricultural Experiment Station
  Department of Agriculture, Ministry of Agriculture
  Sidi Mesri, Tripoli

LIBYA (continued)

- Government Fruit Processing Plant
  P.O. Box 452, Tripoli

- Date Varietal Test Station
  El Guea, Tripolitania

- Date Varietal Test Station
  Hun, Fezzan

MAURITANIA

- Mission IFAC en Mauritanie
  B.P. 87, Nouakchott

- Périmètre fruitier de Rindiao (IFAC)
  B.P. 56, Kaédi

MEXICO

- CIANO (Centro Investigaciones Agrícolas del Noroeste)
  Agricultural Centre for the North West
  Ciudad Obrégon, Sonora

MOROCCO

- Station Centrale des Améliorations Culturales
  B.P. 415, Rabat

- Station Centrale de Phytiatrie
  B.P. 415, Rabat

- Direction de la Recherche agronomique
  Avenue de Temara, Rabat

- Station centrale de Recherches sur le Palmier-Dattier
  B.P. 533, Marrakech

- Station expérimental phoenicicole
  Zagora

NIGER

- Mission IFAC au Niger
  B.P. 886, Niamey

- Station expérimentale (IFAC)
  B.P. Gabougoura

NIGERIA

- Nigerian Institute for Oil Palm Research
  P.M.B. 1030, Benin City

PAKISTAN

- Ayub Agricultural Research Institute
  Lyallpur Faisalabad.

- Government Fruit Research Station
  Mirpurkhas

- Agricultural Research Institute
  Tarnab Farm (Peshawar)

- Date Palm Research Station
  Muzaffargarh (Telari Bag)

- Date Palm Research Station
  Jhang

- Government Date Processing Plant
  Turbat Mekran (Lahore)

SAUDI ARABIA

- Ministry of Agriculture and Water
  Riyadh

- Government Experimental Farm
  Hufuf

SOMALI DEMOCRATIC REPUBLIC

- Ministry of Agriculture
  Mogadisciu

- Date Introduction Station
  Bihindula

SOUTH AFRICA (UNION OF)

- Citrus and Subtropical Fruit Research Institute
  Private Bag, Nelspruit (Eastern Transvaal)

- Fruit and Food Technology Research Institute
  Private Bag, Stellenbosch (Cape Province)

SPAIN

- Instituto Nacional de Investigaciones Agrarias (INIA)
  Avenida Puerta de Hierro, Madrid 3

SUDAN

- Research Division of Ministry of Agriculture
  Hudeiba Research Station, Hudeiba (Northern Province)

- Date and Fruit Factory
  P.O. Box 31, Karim 2

SYRIA

- Horticultural Research Division
  Directorate of Agricultural Scientific Research
  Ministry of Agriculture and Agrarian Reform
  Damascus - Douma

THAILAND

- Northeast Agricultural Centre - Horticultural Branch
  Tha Phra, Khon Kaen

TUNISIA

- Institut national de la Recherche agronomique de Tunisie (INRAT)
  Avenue de l'Independance, Ariana, Tunis

- Service Pédologique
  Avenue de la République, Tunis-Port

TURKEY

- Horticultural Research Institute
  Antalya

UNITED STATES OF AMERICA

- U.S. Date and Citrus Station (USDA)
  44-455 Clinton Street, Indio, California 92201

- Department of Horticultural Science
  University of California, Riverside, California 92502

- Department of Entomology
  University of California, Riverside, California 92502

- Agricultural Engineering Department
  University of California, Riverside, California 92502

UNITED STATES OF AMERICA (continued)

- Date Growers' Institute of the Coachella Valley
  Coachella, California

- Salt River Valley Citrus Branch Station
  Temple, Arizona

- Texas Agricultural Experiment Station (Substation 19)
  Winter Haven, Texas

APPENDIX III

LIST OF REFERENCES

1. ABOU EL-GHAR, M.R. and EL-RAFIE, M.S. 1964. Control measures of date pests in Egypt. Bull. Soc. Entomol. d'Egypte, 48 : 291-301.

2. AFCHAR, J. 1916. Insectes nuisibles aux arbres fruitiers en Iran. Teheran.

3. AHMAD, M. and ALI, N. 1960. Effect of different pollens on the physical and chemical characters and ripening of date fruit. Punjab Fruit J., 23 (80) : 10-11.

4. AHMAD, M.N. and ULLAH, M.H. 1951. Date cultivation in the Punjab (Pakistan). Punjab Fruit J., 15 (50) : 9-14.

5. AL-ANI, H.Y., EL BEHALI, A., MAJEED, H.A. and MAJEED, M. 1971. Reaction of date palm cultivars to inflorescence rot and persistancy and spreading of the disease. Phytopathologia Mediterranea 10 (1) : 57-62. Baghdad University, Abu-Ghraib, Iraq.

6. AL-ANI, H.Y., EL BEHALI, A., MAJEED, H.A. and MAJEED, M. 1971. The control of date palm inflorescence rot. Phytopathologia Mediterranea 10 (1) : 82-85. Baghdad University, Abu-Ghraib, Iraq.

7. AL-BAKIR, A.J. and AL-AZZAWI, J. 1965. Date palm studies at Zaafaraniya. Horticultural Experimental Station. 2nd FAO Tech. Conf. Imp. Date Prod. and Proc. Paper No. 20.

8. ALBERT, D.W. 1925. Effects of soil fertilization on the date palm. Rep. Date Growers' Inst., 2 : 4-6.

9. ALBERT, D.W. 1930. Viability of pollen and receptivity of pistillate flowers. Rep. Date Growers' Inst., 7 : 5-7

10. ALBERT, D.W. and HILGEMAN, R.H. 1932. A study of bud growth in the date palm. Rep. Date Growers' Inst., 9 : 5-6.

11. ALDRICH, W.W. and CRAWFORD, C.L. 1941. Second report upon cold storage of date pollen. Ann. Rep. Date Growers' Inst., 18 : 5.

12. ALDRICH, W.W., FURR, J.R., CRAWFORD, C.L. and MOORE, D.C. 1946. Checking of fruits of the Deglet Noor date in relation to water deficit in the palm. J. Agr. Res., 72 : 211-231.

13. ALDRICH, W.W., LEACH, G.H. and DOLLINS, W.A. 1945. Some factors influencing the growth of date offshoots in the nursery row. Proc. Amer. Soc. Hort. Sci., 46 : 215-221.

14. ALDRICH, W.W., FURR, J.R., CRAWFORD, C.L. and MOORE, D.C. Checking of fruits
1946 of the Deglet Noor date in relation to water deficit in the palm. J. Agric. Res., 72 : 211-231.

15. ALEXANDER, D.B.W. A method of pollinating dates. Ann. Rep. Date Growers'
1952 Inst., (29) : 20.

16. AL-HARDAN, D.S.M. and AL-TAIE, F.H. Soil and water salinity effect on date
1965 palm vigour along the Tigris River. 2nd FAO Tech. Conf. on Imp. Date Prod. and Proc., Paper no. 29.

17. AL-KHAFAF, A.M. Mechanization of palm cultivation and date processing in
1975 Iraq. 3rd FAO Tech. Conf. on Imp. Date Prod., Proc. and Mktg., Paper No. C.4.

18. ALLISON, J.L. Diseases of economic plants in Iraq. FAO Plant Prot. Bull.,
1952 1 : 9-11.

19. AL-SOUSI, A. Stored date insects in Iraq. 2nd FAO Tech. Conf. on Imp. Date
1965 Prod. and Proc., Paper no. 30.

20. ANSELME, C. and BALTZAKIS, N. Sur une pourriture de dattes de conservation
1957 provoquée par Mauginiella scaettae, Cav. Ann. Epiphyt., 8 : 153-164.

21. ARAR, A. Soils, irrigation and drainage of the date palm. 3rd FAO
1975 Tech. Conf. on Imp. Date Prod., Proc. and Mktg., Paper no. A.3.

22. ARMITAGE, H.M. and STEINWEDEN, J.B. The fumigation of California dates with
1945 methyl bromide. Bull. Calif. Dept. Agr., 34 : 101-107.

23. ARMSTRONG, W.W. Jr., and FURR, J.R. Root distribution of Khadrawy date palms
1959 in relation to frequency of irrigation. Ann. Rep. Date Growers' Inst., 36 : 13-15.

24. ARRIGONI DEGLI ODDI, E. Ornitologia Italiana. Hoepli, Milan.
1929

25. ARROW, G.J. Fauna of British India, Part I : Coleoptera, Lamellicornia,
1910 Cetoninae and Dynastinae. London.

26. AYERS, A.D. and HAYWARD, H.E. A method for measuring the effects of soil
1949 salinity on seed germinations with observations on several crop plants. Proc. Soil Sci. Amer., 13 : 224-226.

27. BACON, G.H. Crops of the Sudan. TOTHILL, J.D. Agriculture in the Sudan.
1948 O.U.P., London, pp. 368-386. (Bacon states that most of the information was supplied by BEVAN, T.D.)

28. BAILEY, L.H. The standard encyclopaedia of horticulture. Macmillan, New
1950 York, 3 vols.

29. BAKHSHI, J.C. Propagation of date - removal of offshoots. Punjab Horti-
1972 cultural J., 12 (1) : 58-60. Punjab Agricultural University, Ludhiana, India.

30. BALACHOWSKY, A. 1925 Note sur deux prédateurs du Parlatoria blanchardi, Targ. et sur leur utilisation en vue de la lutte biologique contre ce Coccide. Bull. Soc. Hist. Nat. Afr. N., 16 : 167-172.

31. BALACHOWSKY, A. 1926 Note sur l'acclimatation des prédateurs du Parlatoria blanchardi, Targ. dans la palmeraie de Colomb-Béchar en vue de la lutte biologique contre ces Coccides. Bull. Soc. Hist. Nat. Afr. N., 17 : 93.

32. BALACHOWSKY, A. 1928 Observations biologiques sur les parasites des Coccides du Nord-Africa. Ann. Epiphyt., 14 : 280.

33. BALACHOWSKY, A. 1932 Etude biologique des Coccides du bassin occidental de la Méditerranée. Paris.

34. BARREVELD, W.H. 1963 Survey of the Saudi Arabian date industry. FAO Report No. 05932.

34A. BARREVELD, W.H., SMEAD, P.F. and TOBY, J.G. 1974 Report to the Government of Iran on Date Production, Processing and Marketing. FAO Report Ta. 3269. FAO, Rome.

35. BEDFORD, H.W. 1937 Entomological section report, season 1935-6. Agric. Res. Service Rep., 1936, pp. 38-52. Dep. Agric. Sudan, Wad Medani.

36. BELGRAVE, C.D. 1923 Siwa, the oasis of Jupiter Ammon. 275 pp. illus., Lane, London.

37. BILLINGS, S.C. 1963 Consolidated list of approved common names of insecticides and certain other pesticides. Bull. Entomol. Soc. Am., 9 : 189-197.

38. BLATTER, E. 1926 The palms of British India and Ceylon. Milford, London.

39. BLISS, D.E. 1937 Crosscuts in the fruitstalks of date palms. Ann. Rep. Date Growers' Inst., 14 : 8-11.

40. BLISS, D.E. 1944 Omphalia root rot of the date palm. Hilgardia, 16 : 15-124.

41. BLISS, D.E. and LINDGREN, D.L. 1947 The use of Thiomate "19" on dates, and its effect on fruit spoilage. Ann. Rep. Date Growers' Inst., 24 : 5-9.

42. BLISS, D.E. and MATHEZ, F. 1946 The Arkell date garden fertilizer experiment. Ann. Rep. Date Growers' Inst., 23 : 25-33.

43. BLISS, D.E., LINDGREN, D.L., WILBUR, W.D. and VINCENT, L.E. 1950 Second report on date-bunch covers and their relation to the fruit spoilage complex on Deglet Noor dates. Ann. Rep. Date Growers' Inst., 27 : 7-12.

44. BODENHEIMER, F.S. 1930 Die Schädlingsfauna Palästinas. Mon. ang. Ent., 10, Berlin.

45. BONAVIA, E. 1885 The future of the date palm in India. 118 pp., Thacker, Calcutta.

46. BONDAR, G. 1940 Insectos nocivos e molestias di coqueiro. Bot. Inst. Centr. Fom. Econ., Bahia, No. 8.

47. BONDAR, G. 1943 Notas Entomologicas de Bahia. Rev. Ent. XIV, P. 337.

48. BORDEN, A. 1921 A biological study of the Red Date Palm Scale. J. Agric. Res., XXI, p. 659.

49. BOTTA, P.E. 1841 Relations d'un voyage dans l'Yémen. Paris

50. BOYDEN, B.L. 1929 Progress of Date Scale eradication campaign. Ann. Rep. Date Growers' Inst., 6 : 13-14.

51. BOYDEN, B.L. 1930 Progress of Parlatoria Date Scale eradication. Ann. Rep. Date Growers' Inst., 7 : 16-19.

52. BOYDEN, B.L. 1931 Report of progress: Date Scale eradication. Ann. Rep. Date Growers' Inst., 8 : 12-14.

53. BOYDEN, B.L. 1932 Report of progress: Date Scale eradication. Ann. Rep. Date Growers' Inst., 9 : 9-12.

54. BOYDEN, B.L. 1933 Report of progress: Date Scale eradication. Ann. Rep. Date Growers' Inst., 10 : 10-11.

55. BOYDEN, B.L. 1934 Report of progress: Date Scale eradication. Ann. Rep. Date Growers' Inst., 11 : 11-13.

56. BOYDEN, B.L. 1941 Eradication of the Parlatoria date scale in the United States. USDA misc. Pub. No. 433.

57. BROWN, G.K., PERKINS, R.M., VIS, E.G. 1969 Developing ground level equipment for pollinating dates. Rep. Ann. Date Growers' Inst., Coachella, 46 : 30-34.

58. BROWN, G.K., PERKINS, R.M. and VIS, E.G. 1970 Mechanical pollination experiments with the Deglet Noor date palm in 1969. Ann. Rep. Date Growers' Inst., Coachella, 47 : 19-24.

59. BROWN, J.G. 1922 Date rot. Rep. Ariz. Agric. Exp. Sta., 32 : 606-609.

60. BROWN, J.G. and BUTLER, K.D. 1938 Inflorescence blight of the date palm. J. Agric. Res., 57 : 313-318.

61. BROWN, T.W. 1924 Date palm in Egypt. Tech. and Sci. Services, Hort. Sec. Bull. 43, Min. Agric., Cairo.

62. BROWN, T.W. and BAHGAT, N. 1938 Date palm in Egypt. Min. Agr., Hort. Sec. Booklet No. 24, 117 pp., illus. Cairo.

63. BROWN, W.B. and HEUSER, S.G. Behaviour of fumigants during vacuum fumigation.
1953 I. Penetration of methyl bromide into boxes of dates. J. Sci. Food Agr., 4 : 48-57.

64. BRUN, J. and LAVILLE, E. Observations sur un dépérissement de la couronne
1965 foliaire et du bourgeon terminal du palmier-dattier en République Islamique de Mauritanie. 2nd FAO Tech. Conf. on Imp. Date Prod. and Proc., Paper no. 15.

65. BUHRER, E.M., COOPER, C. and STEINER, G. A list of plants attacked by the
1933 root knot nematode (*Heterodera marioni*). Plant Disease Report, 17 : 64-96.

66. BULIT, J., LOUET, D., BOUHOT, D. and TOUTAIN, G. Recherches sur les fusarioses.
1967 I. Travaux sur le bayoud, fusariose du palmier-dattier en Afrique du Nord. Ann. Epiphyt., 18 : 213-239.

66A. BULIT, J. and GIRARD, M. Traitements des plaies des rejets de palmier-dattier.
1965 2nd FAO Tech. Conf. on Imp. Date Prod. and Proc., Paper no. 12.

67. BUNCE Petermann's Mittheil., p. 223, Perthes, Gotha.
1860

68. BURKILL, I.H. A dictionary of the economic products of the Malay Peninsula.
1935 Crown Agents for the Colonies, London.

69. BUNKNER, P.F. and PERKINS, R.M. Mechanical extraction of date pollen. Ann.
1975 Rep. Date Growers' Inst., 52 : 3-7.

70. BUXTON, P.A. Report on the failure of the date crop of Mesopotamia in 1918.
1918 Agricultural Directorate, M.E.F., Govt. Press, Basra.

71. BUXTON, P.A. Insect pests of the date palm in Mesopotamia and elsewhere.
1920 Bull. Ent. Res., 11 : 287-303.

72. CALCAT, A. Diseases and pests of the date palm in the Sahara and North
1959 Africa. Plant Prot. Bull., 8 (1) : 1-6. Reprinted as Paper No. 10, First FAO Date Meeting, Tripoli, 1959.

73. CALDERON, M., POLACEK, R.K. and SCHULBERG, A. A method for date fumigation
1965 under field conditions in Israel. Pub. Nat. Univ. Inst. Agr., Israel, 5 : 501 (Abstract).

74. CARLSON, T. Report to the Government of Saudi Arabia on the establishment
1954 of date processing and packing plants. FAO/EPTA Report no. 266.

75. CARPENTER, J.B. Root-knot nematode damage to date palm seedlings in relation
1964 to germination and stage of development. Ann. Rep. Date Growers' Inst., 41 : 10-14.

76. CARPENTER, J.B. Note on Bayoud disease of date palms in Algeria. Ann. Rep.
1971 Date Growers' Inst., 48 : 14-15.

77. CARPENTER, J.B. Date palm research and culture in Morocco with special
1973 reference to Bayoud disease. Ann. Rep. Date Growers' Inst., 50 : 11-12.

78. CARPENTER, J.B. 1975 Notes on date culture in the Arab Republic of Egypt, Israel and P.D.R. Yemen. Ann. Rep. Date Growers' Inst., 52 : 18-24.

79. CARPENTER, J.B. and KLOTZ, L.J. 1966 Diseases of the date palm. Ann. Rep. Date Growers' Inst., 43 : 15-21.

80. CATY, R. 1929 Les exigences et les aptitudes du dattier. Acad. Sci. Colon., Paris, 3 : 227-293.

81. CAUVET, S. 1914 La culture du palmier au Souf. Notes prises à El Oued en 1900-1901. Rev. Afric. Nord, 292 (3) : 61.

82. CAVARA, F. 1925a Atrofia fiorale, in Phoenix dactylifera di Cirenaica. Atti R. Accad. di Lincei, Ser., 6 (1) : 65-67.

83. CAVARA, F. 1925b Mauginiella scaettae, Cav. Nuovo ifomicete parassita della palma da datteri di Cirenaica. Boll. Ortobot. Napoli, 8 : 207-211.

84. CHABROLIN, C. 1928 La pourriture de l'inflorescence du palmier-dattier. Ann. Epiphyt., 14 : 377-414.

84A. CHABROLIN, C. 1930 Les maladies du dattier. Rev. Bot. Appl., 10 : 557-566, 661-671.

85. CHAMBERS 1879 (Note on leaf miner). Rep. Dept. Agric. Florida, 210 pages.

86. CHARLET, L. 1905 Les palmiers du M'Zab. Soc. de Géog. d'Alger et de l'Afrique du Nord, Bull. 10 (11) : 87.

87. CHEVALIER, A. 1930 Le dattier en Mauritanie. Rev. Bot. Appl., 10 : 372-376.

88. CHEVALIER, A. 1932 Les productions végétales du Sahara et de ses confins nord et sud, passé - présent - avenir. Rev. Bot. Appl., 12 : 669-924.

89. CIPOLLA, G. 1953 Bayoud: enfermedad de la palmera dattilera. Idia, 1955 : 1 - 24.

90. COMELLY, A. 1960 Le palmier-dattier en Israël. Fruits, 15 : 223-231.

91. COMMONWEALTH INSTITUTE OF ENTOMOLOGY. 1962 Parlatoria blanchardi (Targ.) Distribution maps of pests, Ser. A. (Agricultural), Map 148.

92. COOK, C.E. 1924 Growing and handling date offshoots. Ann. Rep. Date Growers' Inst., 1 : 18, 19.

93. COOK, R.E. 1956 A study of the relationship of heat units to the ripening time of dates. Ann. Rep. Date Growers' Inst., 33 : 13.

94. COOK, W.W. 1941 Economic results of changes in volume and quality of production of dates of Deglet Noor variety. Ann. Rep. Date Growers' Inst., 18 : 27-29.

95. CORBET, A.S. and TAMS, W.H.T. Keys for the identification of the Lepidoptera
1943 infesting stored food products. Proc. Zool. Soc. (London), Ser. B., 113 : 55-148.

96. COSSON, E. Considérations générales sur le Sahara algérien et ses
1859 cultures. Bull. Soc. Imp. Zool. d'Acclimatation, VI, p. 7, Masson, Paris.

97. COSSON, E. Le règne végétal en Algérie. Paris.
1897

98. COSSON, E. and JAMIN, P. De la culture du dattier dans les oasis des Ziban.
1855 Bull. Soc. Bot. de France, II, pp. 36-51. Paris.

99. CRAWFORD, C.L. Cold storage of date pollen. Ann. Rep. Date Growers' Inst.,
1938 15 : 20.

100. CRIDER, F.J. Propagation of the date palm, with particular reference to the
1926 rooting of high offshoots. Agric. Exp. Sta., Univ. of Arizona, Bull. No. 115. Tucson.

101. CROSSA-RAYNAUD, P. Problèmes d'arboriculture fruitière en Tunisie. Ann.
1960 Inst. Nat. Rech. Agron. (Tunisie), 33 : 1-263.

102. CUDEBEC, C.L. OFFSHOOTOLOGY. Ann. Rep. Date Growers' Inst., 6 : 4-7.
1929

103. DA COSTA-LIMA, A. Outras moscas cujas larvas sao predadores de Coccideos.
1937 Characas e Quintais, Sao Paulo.

104. DANTHINE, H. Le palmier-dattier et les arbres sacrés dans l'iconographie
1937 de l'Asie occidentale ancienne. 277 pp., 206 plates, Geuthner. Paris.

105. DARLEY, E.F., NIXON, R.W. and WILBUR, W.D. An unusual disorder of Barhee
1960 date palms. Ann. Rep. Date Growers' Inst., 37 : 10-12.

106. DARLEY, E.F. and WILBUR, W.D. Results of experiments on control of fruit
1955 spoilage of Deglet Noor and Saidy dates in California, 1935 to 1954. Ann. Rep. Date Growers' Inst., 32 : 14-15.

107. DARLEY, E.F. and WILBUR, W.D. Progress report on rhizosis or rapid decline
1951 of the date palm. Ann. Rep. Date Growers' Inst., 28 : 5-8.

108. DARLEY, E.F., NIXON, R.W., WILBUR, W.D. and CARPENTER, J.E. Second report
1964 on the bending of tops of Barhee date palms. Ann. Rep. Date Growers' Inst., 41 : 16.

109. DE CANDOLLE, A.L.P.P. Géographie botanique raisonnée. Paris and Geneva,
1855 2 vols., Vol. I, pp. 342, 371.

110. DE CANDOLLE, A.L.P.P. Origine des plantes cultivées, p. 240.
1883

111. DELASSUS, M. and PASQUIER, R. Les ennemis du dattier et de la datte.
1932 C.R. Gén. Semaine du Dattier, 5-11 Nov. 1931, Alger, pp. 255-277.

112. DE MIRE, P.B. Les insects Rhinoceros du palmier-dattier (Col., Scarabaeidae, trib. Dynastini). J. d'Agric. Trop. et de Bot. Appl., VII, 4-5, 241-255.
1960

113. DESERT LOCUST CONTROL ORGANIZATION FOR EASTERN AFRICA. First annual report
1964 (1st October 1962 to 30th June 1963). Kenya, Government printer, 32p.

114. DE VAJDA, A. Some aspects of surface water developments in arid regions.
1952 FAO. Agr. Devt. Paper No. 21.

115. DEWAN, M.L. and BOOHER, L.J. Soils and irrigation in relation to date palm
1965 production. 2nd FAO Tech. Conf. on Imp. Date Prod. and Proc., Paper No. 4.

116. DILLMAN, R.S. et al. Date management practices at Bard, California. Ann.
1945 Rep. Date Growers' Inst., 22 : 16-17.

117. DONAHAYE, E. and CALDERON, M. Survey of insects infesting dates in storage
1964 in Israel. Israel J. Ag. Res., 14 (3) : 97-100.

118. DONOHOE, H.C. et al. Biology of the raisin moth. USDA Tech. Bull. 994.
1949

119. DOWSON, V.H.W. Dates and date cultivations of Iraq, Part I - The cultivation
1921a of the date palm on the Shatt al-'Arab. Ag. Directorate of Mesopotamia, Memoir III, Heffer, Cambridge.

120. DOWSON, V.H.W. Dates and date cultivation of Iraq. Part II - The results
1921b of an investigation into the yield of date palms on the Shatt al-'Arab. Ag. Directorate of Mesopotamia, Memoir III, Heffer, Cambridge.

121. DOWSON, V.H.W. Dates and date cultivation of Iraq. Part III - The varieties
1923a of date palms in the Shatt al-'Arab. Ag. Directorate of Mesopotamia, Memoir III, Heffer, Cambridge.

122. DOWSON, V.H.W. Dates and date palms in French North Africa. Report to the
1923b Government of Iraq. Baghdad.

123. DOWSON, V.H.W. A serious pest of date palms, Ommatissus binotatus, Fieb., Trop.
1936 Agric., Trin., 13 (7) : 180-181.

124. DOWSON, V.H.W. Date cultivation in the northern Sudan. Report to the
1948 Government of the Sudan. Felden. Type. 135 pp.

125. DOWSON, V.H.W. Report on date scheme. Report to the Government of
1954 Somaliland Protectorate. Djibouti.

126. DOWSON, V.H.W. Report to the Government of Libya on date production.
1961 See FAO.

127. DOWSON, V.H.W. Report to the Government of Somalia on date production.
1963 FAO/EPTA Report no. 1731.

128. DOWSON, V.H.W. Report to the Government of Iran on date production.
1964 FAO/EPTA Report no. 1824.

129. DOWSON, V.H.W. and ATEN, A. Dates: handling, processing and packing. FAO
1962 Agric. Devt. Paper no. 72.

130. DOWSON, V.H.W. Report to the Government of Libya on date production. Rome.
1961 FAO/EPTA Report no. 1263.

131. DOWSON, V.H.W. Bibliography of the date palm. Pub. Field Research Projects,
1976 Coconut Grove, Miami, Florida.

132. DOWSON, V.H.W. and PANSIOT, F.P. Background paper on improvement of date
1965 palm growing. 2nd FAO Tech. Conf. on Imp. Date Prod. and Proc., Paper no. 1.

133. DRUMMOND, B. Propagation and culture of the date palm. Farmers' Bull.
1919 1016. USDA, Washington, D.C.

134. DUNCAN, H.O. Securing higher date yields and improving quality. Ann. Rep.
1941 Date Growers' Inst., 18 : 22-23.

135. DURAND, J.H. and GUYOT, J. L'irrigation des cultures dans l'Oued Rhir.
1955 Trav. de l'Inst. de Rech. Sahariennes, XIII.

136. DURRANT, J.H. Myelois phoenicis, Species Novo, bred from dates in Algeria
1915 and England. Ent. Mon. Mag., (3) LI, 303-304.

137. DUTT, A. Supplementary note on the pests of the date palm in Iraq.
1922 Mesopotamia Dept. Agric., Memoir 6 : 13-21.

138. EAST AFRICAN COMMON SERVICES ORGANIZATION. Report of the desert locust
1962 survey, 1st June 1955 to 21st May 1961. Kenya, Government printer.

139. EAST AFRICAN COMMON SERVICES ORGANIZATION. Final report of the desert locust
1963 survey, 1st June 1961 to 30th September 1962. Kenya, Government printer. 19p.

140. EAST AFRICAN HIGH COMMISSION. Report of the desert locust survey and control,
1953 1st October 1950 to 21st December 1952. Nairobi, Zenith Printing Works Ltd. 64p.

141. EFFRON, H. Marketing and economic aspects of date production and
1975 processing. 3rd FAO Tech. Conf. on Imp. Date Prod., Proc. and Mktg. **Background paper, Item 8.**

142. EL-BAKER, ABDUL-JABBAR. Report to the Government of Saudi Arabia on date
1952 cultivation in Saudi Arabia. FAO/EPTA Report no. 31.

143. ELMER, H.S. 1961 — 1960 results using new insecticides and methods for the control of date insects and mites. Ann. Rep. Date Growers' Inst., 38 : 17-18.

144. ELMER, H.S. 1963 — Protection of dates from injury caused by the Apache cicada in California. J. Econ. Ent., 56 : 875-876.

145. ELMER, H.S. 1964 — Present status of date pest control studies. Ann. Rep. Date Growers' Inst., 41 : 4-6.

146. ELMER, H.S. 1965 — Banks grass mite, *Oligonychus pratensis*, on dates in California. J. Econ. Ent., 58 : 531-534.

147. ELMER, H.S. 1966 — Date palm insect and mite pests in United States. Ann. Rep. Date Growers' Inst., 43 : 9-14.

148. ELMER, H.S. 1967 — Malathion, a factor in date harvesting. Ann. Rep. Date Growers' Inst., 44 : 7-8.

149. EL-WAQIL, A.T. 1965 — Date production in the U.A.R. 2nd FAO Tech. Conf. on Imp. Date Prod. and Proc., Paper no. 14.

150. EMBLETON, T.W. and COOK, J.A. 1947 — The fertilizer value of date leaf and fruit stalk prunings. Ann. Rep. Date Growers' Inst., 24 : 18-19.

151. ESSIG, E.O. 1915 — Injurious and beneficial insects of California. Suppt. to Monthly Bull. Calif. State Comm. Hort. 2nd Ed. Sacramento.

152. EVREINOFF, V.A. 1956 — Contribution à l'étude du dattier. J. Agric. Trop. 3 : 328-333.

153. FAIRCHILD, D.G. 1903 — Persian Gulf dates and their introduction into America. Bur. Plant Indust. Bull. 54, USDA, Washington, D.C.

154. FAIRMAIRE 1859 — *Metholeus cylindricus*, Germ. An. Soc. Ent. Fr. (3) VII, 53.

155. FAO 1976 — Irrigation and Drainage Paper 29 "Water Quality for Agriculture" - R.S. Ayers and D.W. Westcot, FAO, Rome.

156. FAO 1959 — Background paper on date production, by FAO Crop Production and Improvement Branch, Tripoli Libya. 1st FAO Int. Tech. Mtg. on Date Prod. and Proc., Paper no. 1.

156A. FAO 1959 — Background paper on date processing and by-products utilization, by FAO Agricultural Engineering Branch, Tripoli, Libya. 1st FAO Int. Tech. Mtg. on Date Prod. and Proc., Paper no. 2.

157. FAO 1960 — Report of the First FAO International Technical Meeting on Date Production and Processing, Tripoli, Libya, 5-11 December 1959. No. PL/1960/4.

158. FAO 1962-1978 — Production Yearbooks, Vols. XVI to XXXII, FAO, Rome.

159. FAO 1962-1978 Trade Yearbooks, Vols. XVI to XXXII, FAO, Rome.

160. FAO 1965 The date and nutrition, by Nutrition Division, FAO, Rome (English). Baghdad. 2nd FAO Tech. Conf. on Imp. Date Prod. and Proc., Paper no. 2.

161. FAO 1966 Report of the Second FAO Technical Conference on the Improvement of Date Production and Processing. Held in Baghdad, Iraq, 16-25 October 1965. Meeting Report No. PL/1965/16.

162. FAO 1976 Report of the Third FAO Technical Conference on the Improvement of Date Production, Processing and Marketing. Held in Baghdad, Iraq, 30 November-4 December 1975. AGP:1976/M/2.

163. FARIES, W.R. 1924 The rooting of high offshoots on the palm. Ann. Rep. Date Growers' Inst., 1 : 20.

164. FAWCETT, H.S. 1930 An offshoot and leafstalk disease of date palms due to Diplodia. Phytopathology, 20 : 339-344.

165. FAWCETT, H.S. 1931 Observations on the culture and diseases of date palms in North Africa. Ann. Rep. Date Growers' Inst., 8 : 18-23.

166. FAWCETT, H.S. and KLOTZ, L.J. 1932 Diseases of the date palm. Phoenix dactylifera. Bull. Calif. Agric. Exp. Sta., 522, 47p.

167. FENZI, E.O. 1916 Frutti tropicali e semitropicali. Biblioteca Agraria Coloniale, Istituto Ag. Col. It., Florence.

168. FISCHER, E. 1883 Beitrag zur Kentniss der Gattung Graphiola. Bot. Zeit., 41 : 745-756, 761-773, 777-788, 793-801.

168A. FISCHER, T. 1881 Die Dattelpalme, ihre geographische Verbreitung und culturhistorische Bedeutung. Ergänzungschaft Nc. 64 zu Petermanns Mittheilungen, pp. 85. Perthes, Gotha.

169. FORSKÅL, P. 1775 Descriptionae animalium quae in itinere orientali observavit. Copenhagen, p. 85.

170. FREEMAN, G.F. 1910 Dates. Ann. Rep. Ariz. Agric. Exp. Sta., 21 : 384-385.

171. FREEMAN, G.F. and BRYAN, W.E. 1917 Dates. Ann. Rep. Ariz. Agric. Exp. Sta., 28 : 455.

172. FURR, J.R. 1975 Water and salinity problems of Abadan island date gardens. Ann. Rep. Date Growers' Inst., 52 : 14-17.

173. FURR, J.R. and ARMSTRONG, W.W. Jr. 1957 Nitrogen fertilization of dates - a review and progress report. Ann. Rep. Date Growers' Inst., 34 : 6-9.

174. FURR, J.R. and ARMSTRONG, W.W. Jr. 1959 The relation of growth, yield and fruit quality of Deglet Noor dates to variations in water and nitrogen supply and to salt accumulation in the soil. Ann. Rep. Date Growers' Inst., 36 : 16-18.

175. FURR, J.R. and ARMSTRONG, W.W. Jr. A test of mature Halawy and Medjool date
1962 palms for salt tolerance. Ann. Rep. Date Growers' Inst., 39 : 11-13.

176. FURR, J.R., ARMSTRONG, W.W. Jr. and LIEU, P.S. Fruit quality in relation to
1961 crop load of Deglet Noor dates. Ann. Rep. Date Growers' Inst., 38 : 4-6.

177. FURR, J.R. and BARBER, H.D. The nitrogen content of some date garden soils
1950 in relation to soil management practices. Ann. Rep. Date Growers' Inst., 27 : 26-30.

178. FURR, J.R. and BROWN, T.R. Comparative yields of Deglet Noor palms fertilized
1963 with manure or ammonium nitrate. Ann. Rep. Date Growers' Inst., 40 : 2-3.

179. FURR, J.R., CURRLIN, E.C., HILGEMAN, R.N. and REUTHER, W. An irrigation and
1952 fertilization experiment with Deglet Noor dates. Ann. Rep. Date Growers' Inst., 29 : 10-12.

180. GANDZ, S. Artificial fertilization of date palms in Palestine and Arabia.
1935 Isis, 23 : 245-250.

181. GAZALA, C. and BEY, S. Contribution à la parasitologie des plantes cultivées
1931 en Irak. Thèse Université de Nancy. 140 pp., 12 plates, "Le dattier, troisième partie", pp. 69-112.

182. GERARD, B. The effect of heat on the germination of date pollen. Ann.
1932 Rep. Date Growers' Inst., 9 : 15.

183. GHOSH, C.C. Life-histories of insect pests, III. Memoirs of the Dep.
1911 Agric. in India. Entomological Series, 2 (10). Govt. Printers, Calcutta.

184. GOOR, A. The history of the date through the ages in the Holy Land.
1967 Econ. Bot., 21 : 320-340.

185. GRASOVSKY, A. and WAITZ, J. The date palm in Palestine. Hadar, V 8-9
1932 (reprinted as Ag. Leaflet No. 29, Hort. Sec., Dep. Agric. and Forests, Jaffa).

186. GRIDLEY, T.J. Management of a bearing date garden. Ann. Rep. Date Growers'
1924 Inst., 1 : 7-9.

187. GAUTIER, Lt.-Col. L'Anarchie hydraulique dans le territoire de Touggourt.
1935 Compte-rendu général des journées du dattier, Bisra-Touggourt, 13-17, 11 (33) : 25-46. OFALAC, Algiers.

188. HAAS, A.R.C. Mineral nutrition of the date palm. Ann. Rep. Date Growers'
1930 Inst., 7 : 19-21.

189. HAAS, A.R.C. Inorganic composition of date fruits. Ann. Rep. Date Growers'
1935 Inst., 12 : 5-8.

190. HAAS, A.R.C. 1944 Boron in the palms and soil of date gardens in the Coachella Valley of Southern California. Proc. Amer. Soc. Hort. Sci., 44 : 34-42.

191. HAAS, A.R.C. and BLISS, D.E. 1935 Growth and composition of Deglet Noor dates in relation to water injury. Hilgardia, 9 : 295-344.

192. HAECKEL, E. 1883 A visit to Ceylon. (Trans. from the German by Clara Bell). London.

193. HALL, W. 1929 Observations on the Coccidae of Southern Rhodesia. Bull. Ent. Res., 19 : 271.

194. HARRIS, K. 1955 Report on the soils and other agricultural conditions of the Qatif area (being exhibit no. 3 of Knapp's Report). Dhahran. (mimeo).

195. HARRISON, P.W. 1940 Doctor in Arabia. 303 pp., illus. John Day, New York.

196. HAYES, W.B. 1953 Fruit growing in India. 2nd Ed. Allahabad, Kitabistan. 449p.

197. HAYWARD, H.E. 1949 The salt problem and the salt tolerance of irrigated crops. Ann. Rep. Date Growers' Inst., 26 : 15-17.

198. HAYWARD, B. 1925 Fertilization of date palms. Ann. Rep. Date Growers' Inst., 2 : 819.

199. HEDLUND, F.F. 1965 Date production in the United States. 2nd FAO Tech. Conf. on Imp. Date Prod. and Proc., Paper no. 23.

200. HELGESON, E.A. 1957 Methods of weed control. FAO Agric. Study no. 36.

201. HILGEMAN, R.H. 1953 Report to the Government of Israel on the culture of citrus fruits and the date palm. FAO/EPTA Report no. 100.

202. HODGSON, F.M. 1942 Persian Gulf Pilot. Hydrographic Dept., the Admiralty, London.

203. HODGSON, R.W. 1925 The fertilization of fruit trees. Ann. Rep. Date Growers' Inst., 2 : 1-4.

204. HODGSON, R.W. 1932 Date culture in Tunisia - miscellaneous observations elsewhere in the Mediterranean. Ann. Rep. Date Growers' Inst., 9 : 7-12.

205. HOUSER 1918 *Aonidiella orientalis*, Newst. (=*Aspidiotus cocotiphagus*, Marlatt.) on coconut palm and date palm. An. Ent. Soc. Amer., XI, 157.

206. HUGHES-BULLER, R. 1906 Makran: Baluchistan District Gazetteer Series, Vol. VII. The Times Press, Bombay.

207. HUSSAIN, A.A. 1963 Biology and control of the dubas bug [Ommatissus binotutus lybicus de Berg. (Homoptera: Tropiduchidae)] infesting date palms in Iraq. Bull. Entomol. Res., 53 : 737-745.

207A. HUSSAIN, A.A. 1963 Notes on borers of date palms in Iraq. Bull. Entomol. Res., 54 : 345-348.

208. HUSSAIN, A.A. 1975 Insects, mites and diseases of date palms and stored dates. 3rd FAO Tech. Conf. on Imp. Date Prod., Proc. and Mktg., Paper no. A.4.

209. HUSSAIN, A.A. 1965 Notes on dates and date palms in Iraq. 2nd FAO Tech. Conf. on Imp. Date Prod. and Proc., Paper no. 19.

210. HUSSAIN, F. 1958 Occurrence of date palm inflorescence rot in Iraq. Plant Disease Reporter, 42 : 555.

211. IBRAHIM, I.M. 1975 The future of date palm varieties imported to Egypt from Iraq and Saudi Arabia. 3rd FAO Tech. Conf. on Imp. Date Prod., Proc. and Mktg., Paper no. C.33.

212. IGNATIEFF, V. and PAGE, H.J. 1958 Efficient use of fertilizers. FAO Agric. Study no. 43.

213. IMMS, A.D. 1951 A general textbook of entomology. Methuen, London.

214. IRAQ METEOROLOGICAL DEPARTMENT 1919 Climate and weather of Iraq. Govt. Press, Baghdad.

215. IRAQ 1939 onwards Statistical abstracts. Principal Bureau of Statistics. Govt. Press, Baghdad.

216. ISSA BEY, A. 1930 Dictionnaire des noms des plantes. Min. Publ. Inst., Egypt. Imp. Nat., Cairo.

217. JACOB, A. and VON UEXKÜLL, H. 1958 Fertilizer use. Verlagsgesellschaft für Ackerbau mbH. Hannover.

218. JARVIS, E.L. and JARVIS, E.C. 1941 Date production methods. Ann. Rep. Date Growers' Inst., 18 : 23.

219. JENSEN, H.J. 1961 The nematode menace. American Nurseryman, 114 (8) : 7-8, 105-113.

220. JOHNSTON, J.C. 1944 The value and use of fertilizers. Ann. Rep. Date Growers' Inst., 21 : 11-13.

221. JORET, C. 1897 Les plantes dans l'antiquité et au moyen âge. 504 pp., Librairie Emile, Bouillon.

222. JUNAID, A.H.M. and NASIR, M.M. 1956 Fumigation of cut-flowers, rose cuttings and preserved dates with methyl bromide. Agr. Pakistan, 6 (2) : 80-103. (Abstract)

223. JUS, H. 1900 Notice sur l'hydraulique agricole en Algérie. Giralt, Alger.

224. KAEMPFER, E. 1712 Amoenitatum exoticarum, politico-physico-medicarum. Lemgo. Vol. 5 (3), p. 685-686.

225. KAPLAN, J., LENORMAN, C. and COMBA, P. 1972 La protection des régimes de dattier contre les attaques aviaires. Fruits, 27 (6) : 439-444. IFAC, Nouakchott, Mauritania.

226. KEARNEY, T.H. 1906 Date varieties and date culture in Tunis. U.S. Dept. Agr., Bur. of Plant Industry Bull. 92, 112p. Washington, D.C.

227. KEHAT, M., STOLLER, S., MEISNER, J. and GREEMBERG, S. 1966 Development of population and control of Carpophilus dimidiatus F., hemipterus L. and Coccotrypes dactyliperda F. in dates. Israel J. Agric. Res., 16 : 173-176.

228. KEHAT, M. and SWIRSKI, E. 1964 Chemical control of the date palm scale, Parlatoria blanchardi, and the effect of some insecticides on the lady beetle Pharoscymnus aff. numidicus Pic. Israel J. Agric. Res., 14 : 101-110.

229. KENKNIGHT, G. 1948 Findings of Omphalia date root rot survey. Ann. Rep. Date Growers' Inst., 25 : 5-11.

230. KENKNIGHT, G. and AMLING, R.O. 1947 Progress report on the Omphalia date root rot survey. Ann. Rep. Date Growers' Inst., 24 : 10-17.

231. KHUDAIRI, A.K. 1958 Studies on the germination of date palm seeds. The effect of sodium chloride. Physiol Plantarum, (11) : 16-22. Copenhagen.

232. KILLIAN, C. 1924 Le développement du Graphiola phoenicis. Poit. et ses affinités. Rev. Gén. Bot., 36 : 385-394, 451-460.

233. KLEINE, E. 1935 Die Borkenkäfer (Ipidae) und ihre Standpflanzen. Eine vergleichen der Studie. Zeits. Ang. Ent., XXI : 123, 597.

234. KLOTZ, L.J. and FAWCETT, H.S. 1932 Black scorch of the date palm caused by Thieloviopsis paradoxa. J. Agric. Res., 44 : 155-166.

235. KURIAN, C. and PILLAI, G.B. 1964 Rhinocerous beetle. World Crops, 16, Mar., 20-24.

236. LABORATOIRE D'ARBORICULTURE FRUITIERE DE TUNISIE. 1959 Le problème des plantations de palmier-dattier dans la zone côtière en Tunisie. 1st FAO Date Meeting, Tripoli, Libya, 1959. Paper no. 6.

237. LAFFOON, J.L. 1960 Common names of insects. Bull. Entomol. Soc. Amer., 6 (4) : 175-211.

238. LAFLIN, B.T. 1963 Experience in conversion to non-tillage and weed control with oil. Ann. Rep. Date Growers' Inst., 40 : 7-8.

239. LANGRONIER, C. 1935 Besoin en eau du palmier-dattier au M'Zab. Compte-rendu général des Journées du Dattier, Biskra-Touggourt, 13-17, XI. 33, p. 231-233. OFALAC, Algiers.

240. LANGRONIER, C. and MONCIERO, A. 1922 La culture potagère dans les oasis. Doc. et Rens. Agric., Bul. no. 177. Insp. Gén. et Dir. de l'Agric., Algiers.

241. LASSERRE, A. 1922 Aperçu météorologique. Les territoires du Sud de l'Algérie. Carbonel, Algiers.

242. LAVILLE, E. 1962 Infestations expérimentale de jeunes plantules de palmier-dattier par Fusarium oxysporum var. albedinis. Fruits, 17 : 88-89.

243. LAVILLE, E. 1966 Le palmier-dattier en Iraq (agronomie et commerce). Fruits d'Outre-Mer, 21 : 211-220.

244. LAVILLE, E. and LOSSOIS, P. 1963 Méthode de Van der Planck et mode de propagation du bayoud. Observations préliminaires. Fruits d'Outre-Mer, 18 : 249-253.

245. LAVILLE, E. and SACHS, G. 1967 Premières observations sur un grave dépérissement d'origine inconnue du palmier-dattier. Fruits d'Outre-Mer, 22 : 309-312.

245A. LEDING 1928 Determination of length of time during which the flowers of the date palm remain receptive to fertilization. J. Agric. Res., 36 (2) : 129-134.

246. LEFEVRE, F. 1956 La lutte contre les termites à la Station IFAC du palmier-dattier de Kankossa (Mauritanie). Fruits d'Outre-Mer, 11 : 47-50.

247. LEFEVRE, F. 1962 Multiplication du palmier-dattier à la station de Kankossa (République islamique de Mauritanie). Fruits, 17 : 129-131. Paris.

248. LEGRAIN, L. 1929 The boudoir of Queen Shubad. Museum J., 20 : 211-245.

249. LEGROS, J. 1932 Date palm growing in southern Algeria. Mon. Bull. Agric. Sci. Pract., 23 (3) : 82-89.

250. LEPESME, P. 1963 Les insectes des palmiers. Lechevalier, Paris.

251. LEPIGRE, A. 1963 Essais de lutte sur l'arbre contre la pyrale des dattes (Myelois ceratoniae Zeller-Pyralidae). Ann. Epiphyt., 14 : 85-101.

252. LESNE, P. 1900 An. Soc. Ent. Belg., 69 : 611.

253. LESNE, P. 1925 Les coléoptères bostrychides de l'Afrique tropicale française. Paris.

254. LETHIELEEUX, J. 1948 Le Fezzan, ses jardins, ses palmiers. Bascone et Muscat, Tunis. PP. 201-202.

255. LETURCQ, P. 1965 Note sur les travaux de la station expérimentale du palmier-dattier de Kankossa, République Islamique de Mauritanie. 2nd FAO Tech. Conf. on Imp. Date Prod. and Proc., Paper no. 17.

256. LEVANDOWSKY, D.W. 1959 Propagation of clonal Hevea brasiliensis by cuttings. Trop. Agric., 36 : 247-257. Tunis.

257. LINDGREN, D.L., BLISS, D.E. and BARNES, D.F. 1948 Insect infestation of dates, its control and relation to fungus spoilage. Ann. Rep. Date Growers' Inst., 25 : 12-17.

258. LINDGREN, D.L. and VINCENT, L.E. 1953 Nitidulid beetles infesting California dates. Hilgardia, 22 : 97-118.

259. LOUVET, J. and BULIT, J. 1965 Recherches sur la fusariose du palmier-dattier. 2nd FAO Tech. Conf. on Imp. Date Prod. and Proc., Paper no. 11.

260. MAIRE, R. 1932 Pollination, hybridation, sélection appliqué à la culture améliorée du dattier, pp. 250-254 in Semaine du Dattier, 5 au 11 novembre 1931, 764 pp. Algiers.

261. MAIRE, R. 1935 La défense des palmeraies contre le bayoud et le belâat. C.R. Gén. Journées du Dattier, 13-17 novembre 1933, Biskra-Touggourt, Alger. P. 82-93.

262. MAIRE, R. and MELENÇON, G. 1933 Le Belâat, nouvelle maladie du dattier dans le Sahara algérien. C.R. Acad. Sci., Paris, 196 : 1567-1569.

263. MALENÇON, G. 1941 L'infection florale du dattier par le Fusarium albedinis (Kill. et Maire) Mlçn. C.R. Acad. Sci., Paris, 223 : 923-925.

264. MALENÇON, G. 1949 Le bayoud et la reproduction expérimentale de ses lésions chez le palmier-dattier. Mém. Soc. Hist. Nat. Afrique du Nord, Hors. Sér., 2 : 217-228.

265. MALENÇON, G. 1950 La diffusion et l'épidémiologie de la maladie fusarienne du palmier-dattier en Afrique du Nord. Rev. Mycologie, Suppl. Colon., 15 (1) : 45-60.

266. MALLAMAIRE, A. 1950 La rouille du palmier-dattier en Afrique Occidentale française. C.R. Conf. Internl. des Africanistes de l'Ouest, (1945), 1 (1) : 516-517.

267. MARTIN, H.E. 1958 Pests and diseases of date palm in Libya. FAO Plant Prot. Bull., 6 (8) : 120-123.

268. MARTIN, H.E. 1959 Pests and diseases of date palm in Libya. 1st FAO Tech. Conf. Imp. Date Prod. and Proc., Paper no. 9.

269. MASON, S.C. 1908 Date growing in southern California. Official report of the 24th Fruit Growers' Convention of the State of California, pp. 170-178. Sacramento.

270. MASON, S.C. 1925a Date culture in the Sudan. Department of Agriculture and Forests, Khartoum. McCorquodale, London.

271. MASON, S.C. 1925b The minimum temperature for growth of the date palm and the absence of a resting period. J. Agric. Res., 31 : 401-414.

272. MASON, S.C. 1925c Partial thermostasy of the growth centre of the date palm. J. Agric. Res., 31 : 415-453.

273. MASON, S.C. 1927 Date culture in Egypt and the Sudan. USDA Dep. Bull. 1457. Washington, D.C.

274. MASON, S.C. 1930 A sectorial mutation of a Deglet Noor date palm. J. Heredity, 21 : 157-163.

275. McGREGOR, E.A. 1939 The specific identity of the American date mite; description of two new species of *Paratetranychus*. Wash. Entomol. Soc. Proc., 41 : 247-256.

276. MEYRICK, E. 1921 Exotic microlepidoptera, 2 : 451. London.

277. MILNE, D. 1918 The date palm (*Phoenix dactylifera*) and its cultivation in the Punjab. Punjab Govt. Lahore, Thacker, Spink and Co. Calcutta.

278. MINZ, G. 1958 Two *Diplodia* species in date palm. Ktavim, 8 : 213-216.

279. MINZ, G. *et al.* 1963 Plant parasitic nematodes in Israel and their control. (In Hebrew). English summary. Sifriat Hassadeh, Tel Aviv. 84 pages.

280. MOHAMED, I.I. and EL-HAIDERI, H.S. 1965 Mites attacking date palms in Iraq. 2nd FAO Tech. Conf. on Imp. Date Prod. and Proc., Paper no. 31.

281. MOLENAAR, A. 1959 Water lifting devices for irrigation. FAO Agric. Devt. Paper no. 60.

282. MOLENAAR, A. 1960 Irrigation by sprinkling. FAO Agric. Devt. Paper no. 65.

283. MONCIERO, A. 1947 Etude comparée sommaire des différents types de culture du palmier-dattier en Algérie. Fruits, 2 : 374-382.

284. MONCIERO, A. 1949 Rapport sur l'orientation de l'agriculture et le problème expérimental en agronomie Saharienne. Compte rendu Congrès de la Production Dattière, Touggourt, 4-6 april. (9 pp. mimeo).

285. MONCIERO, A. 1950 — La fécondation mécanique du palmier-dattier. Bull. d'Information, 4 (38-1) : Office Tunisien de Standardisation, Tunis. (Congrès International de la Datte, Tunis, 4-12 novembre 1950).

286. MONCIERO, A. 1952 — Report to the Government of Libya on agriculture. Dates. FAO/EPTA Report no. 21 : 121-126.

287. MONCIERO. A. 1954 — Notes sur le palmier-dattier. Ann. Inst. Agric. Algér., 8 : 3-28.

288. MONCIERO, A. 1959 — Le palmier-dattier au Sahara. 1st FAO Int. Tech. Mtg. on Date Prod. and Proc. Tripoli, Libya. Paper no. 3.

289. MONTEITH, J.L. 1965 — Radiation and crops. Experimental agriculture, 1 : 4; 241-251. London.

290. MOORE, W.O. 1958 — A dictionary of geography. Penguin, London.

291. MUIRHEAD, F. — Southern Spain and Portugal. The Blue Guides, MacMillan, London. 337 pages.

292. MUNIER, P. 1951 — Contribution à la mise en valeur du Sahara Soudanais Française. Fruits, 6 : 104-108.

293. MUNIER, P. 1952 — L'Assaba. Essai monographique. Etudes Mauritaniennes no. 3. 72 pages. Centre IFAN (Mauritanie). Saint-Louis, Sénégal.

294. MUNIER, P. 1955a — Le palmier-dattier en Mauritanie. Annales Inst. Fruits et Agrumes Colon, no. 12. 66 pages.

295. MUNIER, P. 1955b — Le palmier-dattier et sa culture en Afrique tropicale française. Inst. Franç. Rech. Fruits. (second edition in 1957).

296. MUNIER, P. 1956 — Possibilités et avenir de l'extension de la culture du palmier-dattier dans les confins Saharo-Scudanais d'A.O.F. et d'A.E.F. Fruits, 11 : 162-167.

297. MUNIER, P. 1957 — Le palmier-dattier en Espagne continentale. Fruits, 12 : 269-279.

298. MUNIER, P. 1959a — Prospection phénicicole des territoires au nord de la République Tchadienne. Inst. Franç. Rech. Fruits Outre-Mer. Paris. 79 pages.

299. MUNIER, P. 1959b — Contribution à l'étude de la phytoclimatologie du palmier-dattier dans le secteur ouest de l'Afrique tropicale nord. 1st FAO Int. Tech. Mtg on Date Prod. and Proc. Tripoli, Libya. Paper no. 7.

300. MUNIER, P. 1960 — Le palmier-dattier et sa culture en Afrique tropicale. Inst. Franç. de Rech. Fruits Outre-Mer. Paris.

301. MUNIER, P. 1963a — Etude des possibilités de la culture du palmier-dattier en République du Mali. Inst. Franç. de Rech. Fruits Outre-Mer. Paris. 79 pages.

302. MUNIER, P. 1963b — Prospection phénicicole du territoire de la République du Niver. Inst. Franç. de Rech. Fruits Outre-Mer. Paris. 38 pages.

303. MUNIER, P. 1972 — Le palmier-dattier en Egypte. Fruits, 27 (11) : 801-805.

304. MUNIER, P. 1975 — Date varieties. 3rd FAO Tech. Conf. on Imp. Date Prod., Proc. and Mktg. Paper no. A.1.

305. MURRAY, J.A.H. *et al.* 1933 — The Oxford English dictionary. Clarendon, Oxford. (Vol. 3, p. 42).

306. N'DIAYE, A.M. and TOURNEUR, J.C. 1972 — Essai de fréquence des applications de soufre contre le Taka (*Olygonychus afrasiatycus* McGr.) dans le Tagant mauritanien. Fruits, 27 (6) : 465-467.

307. NEWSTEAD, R. 1917 — Observations on African scale insects (*Coccidae*). Bull. Ent. Res., 8 : 125-134. London.

308. NICOL, A.A. 1935 — A study of the fig beetle, *Cotinis texana* Casey. Tech. Bull. Ariz. Agric. Exp. Sta., 55 : 157-198.

309. NIEBUHR, C. 1774 — Description de l'Arabie. Baalde, Amsterdam.

310. NIXON, G.E.J. 1951 — The association of ants with aphids and coccids. Commonwealth Inst. Ent., London. 36 pages.

311. NIXON, R.W. 1926 — Experiments with selected pollens. Ann. Rep. Date Growers' Inst., 3 : 11-14.

312. NIXON, R.W. 1928 — The direct effect of pollen on the fruit of the date palm. J. Agric. Res., 36 : 97-128.

313. NIXON, R.W. 1931 — The commercial utilization of differences in times of ripening of dates due to pollen. Ann. Rep. Date Growers' Inst., 7 : 4-5.

314. NIXON, R.W. 1932 — Observations on the occurrence of blacknose. Ann. Rep. Date Growers' Inst., 9 : 3, 4.

315. NIXON, R.W. 1933 — Notes on rain damage to varieties at the U.S. Experimental Date Garden. Ann. Rep. Date Growers' Inst., 10 : 13-14.

316. NIXON, R.W. 1934 — The Dairee date - a promising Mesopotamian variety for testing in the Southwest. USDA Circ. no. 33, 12 pages, illus.

317. NIXON, R.W. 1935 — Metaxenia in dates. Proc. Amer. Soc. Hort. Sci., 32 : 221-226.

318. NIXON, R.W. 1936 — Metaxenia and interspecific pollinations in Phoenix. Proc. Amer. Soc. Hort. Sci., 33 : 21-26.

319. NIXON, R.W. 1937 — The freeze of January 1937 - a discussion. Ann. Rep. Date Growers' Inst., 14 : 19-23.

320. NIXON, R.W. 1942 — Rain and high humidity tolerance of commercial date varieties. Ann. Rep. Date Growers' Inst., 19 : 12-13.

321. NIXON, R.W. 1945 — The need for a monograph of the date varieties of the world. Chronica botanica, 9 : 153-154.

322. NIXON, R.W. 1946 — Bunch protection of the Khadrawy date in relation to sunburn and fruit shrivel. Ann. Rep. Date Growers' Inst., 23 : 10-12.

323. NIXON, R.W. 1950a — Contribution to symposium on 1949-50 frost damage to date palms held under the chairmanship of SWINGLE. Ann. Rep. Date Growers' Inst., 27 : 33-35.

324. NIXON, R.W. 1950b — Date culture in French North Africa and Spain. Ann. Rep. Date Growers' Inst., 27 : 15-21.

325. NIXON, R.W. 1950c — Imported varieties of dates in the United States. USDA Circ. no. 834, 144 pages, illus.

326. NIXON, R.W. 1952 — Ecological studies of date varieties in French North Africa. Ecology, 33 : 215-225.

327. NIXON, R.W. 1953a — Off-type palms in commercial date gardens. Ann. Rep. Date Growers' Inst., 30 : 8-9.

328. NIXON, R.W. 1953b — North America's oldest date gardens. Pacific Discovery, VI (1) : 18-24.

329. NIXON, R.W. 1954 — Date culture in Saudi Arabia. Ann. Rep. Date Growers' Inst., 31 : 15-20.

330. NIXON, R.W. 1957 — Differences among varieties of the date palm in tolerance of Graphiola leaf spot. Plant Disease Reporter, 41 : 1026-1028.

331. NIXON, R.W. 1959 — Effects of gibberellin on fruit stalks and fruit of date palm. Ann. Rep. Date Growers' Inst., 36 : 5-7.

332. NIXON, R.W. 1960 — Observations on date culture in Libya and Tunisia. Ann. Rep. Date Growers' Inst., 37 : 19-22.

333. NIXON, R.W. 1966 — Growing dates in the United States. Agric. Inf. Bull., USDA no. 207, 56 pages. (1st edition 1959).

334. NIXON, R.W. 1969 — Date culture in the Sudan. Ann. Rep. Date Growers' Inst., 44 : 9-14.

335. NIXON, R.W. and CRAWFORD, C.L. Fruit thinning experiments with Deglet Noor
1936 dates. Proc. Amer. Soc. Hort. Sci., 34 : 107-115.

336. NIXON, R.W. and FURR, J.R. Quality of Deglet Noor dates in relation to
1961 frequency of picking. Ann. Rep. Date Growers' Inst., 38 : 14, 15.

337. NIXON, R.W. and FURR, J.R. Frequency of picking in relation to quality of
1962 Deglet Noor dates in 1961. Ann. Rep. Date Growers' Inst., 39 : 16-18.

338. NIXON, R.W. and FURR, J.R. Problems and progress in date breeding. Ann.
1965 Rep. Date Growers' Inst., 42 : 2-5.

339. OCHSE, J.J. et al. Tropical and subtropical agriculture. MacMillan, New
1961 York.

340. OPPENHEIMER, G. and REUVENI, O. Investigation into the causes and
1965 possible correction of disturbed date palm fruit set in the northern Negev. The Volcani Inst. of Agric. Res., Dept. of Hort., Rehovot, 35 pages. Hebrew summary.

341. OSMAN, A.M.A., REUTHER, W. and ERICKSON, L.C. Xenia and Metaxenia studies
1974 in the date palm. Ann. Rep. Date Growers' Inst., 51 : 6-17.

342. OVERHOLTZER, S.D. Securing higher date yields and improving quality. Ann.
1941 Rep. Date Growers' Inst., 18 : 22.

343. PAGLIANO, T. Les ennemis des vergers, des olivettes et des palmeraies.
1951 2e éd. Office Exp. et Vulgarisation Agr., Soc. Ed. Fr. Afr. du Nord. 366 pages.

344. PANSIOT, F.P. et al. Report of the FAO mission to Yemen. Rome. Mimeo.
1960 No. C-1792.

345. PAUL, E.C.S. The possibility of producing hardboard commercially from
1965 date palm fronds, etc. Report to the Industrial Bank of Iraq. Undevpro, Baghdad, Cyclo., 46 pages.

346. PEIGHTAL, W.J. Remarks on the mechanization of date harvesting. Ann. Rep.
1962 Date Growers' Inst., 39 : 18.

347. PEMBERTON Hawaii Plant Rec. XXXVIII, 4 : 168.

348. PEREAU-LEROY, P. Le premier congrès international de la datte. Fruits,
1951 6, 1, 27-28.

349. PEREAU-LEROY, P. Recherches sur la fusariose du palmier-dattier. Ann.
1954 Inst. Fruits Agrumes Colon., no. 8, 27 pages.

350. PEREAU-LEROY, P. Recherches d'un test de sensibilité des variétés de
1957 palmier-dattier à la fusariose. Fruits, 12 : 53-56.

351. PEREAU-LEROY, P. Le palmier-dattier au Maroc, IFAC, Paris, et Minist.
1958 d'Agric., 142 pages. Rabat, Maroc.

352. PERKINS, R.M. and BROWN, G.K. Progress in mechanization of date harvesting.
1964 Ann. Rep. Date Growers' Inst., 41 : 19-23.

353. PERKINS, R.M. and BROWN, G.K. Date harvest mechanization. Calif. Agric.,
1966 20 (2) : 8-10.

354. PERKINS, R.M. and BURKNER, P.F. Mechanical pollination of date palms.
1973 Ann. Rep. Date Growers' Inst., 50 : 4-6.

355. PLINY, C. (The Elder) Historia naturale, Book XIII, cap. iii, 3 columns on
1489 (Trans.) the *Palmae*. Translated into Italian by Cristoforo Landino Fiorentino and published by Bartolamio de Zani de Portesio. Venice.

356. POPENHOE, P.B. Date growing in the Old and New Worlds. West India Gardens.
1913 Altadena, California.

357. POPENHOE, P.B. The pollination of the date palm. J. Amer. Oriental Soc.,
1923 42 : 343-354.

358. POPENHOE, P.B. The distribution of the date palm. The Geographical Review.
1926 New York, 16 (1) : 117-121.

359. POSTLETHWAITE, R.H. *Melilotus indica* as a cover crop. Ann. Rep. Date
1925 Growers' Inst., 2 : 78.

360. POWELL, D. The biology of *Cephalonomia tarsalis*. Ann. Ent. Soc. Amer.,
1938 31 : 44-49.

361. POWELL, W.I. Report on a survey of existing and potential date-growing
1964 areas in the provinces of Nigiurtein, Hargeisa and Burao. Hargeisa, typed, 26 pages.

362. PRESTON, R.D. Pollinating dates by airplane. Ann. Rep. Date Growers' Inst.,
1964 41 : 24.

363. PRITCHARD, A.E. and BAKER, E.W. A revision of the spider mite family
1955 *Tetranychidae*. Pacific Coast Ent. Soc., Memoirs Series, Vol. 2.

364. PRUESSNER, A.R. Date culture in ancient Babylonia. Amer. J. Semitic
1920 Languages and Literatures, Chicago, 36 : 213-232.

365. RAHIM, A.L., AL-GABOURI, N., KALID NOMAN and SAFAA DHIA. Studies on date
1975 palm pollination. 3rd FAO Tech. Conf. on Imp. Date Prod., Proc. and Mktg. Paper no. C.29.

366. RAHMAN and WAHID ABDUL. (Note on date moths). Proc. Int. Acad. Sci., B,
1944 XIX, page 1.

367. RAINEY, R.C. The use of insecticides against the desert locust. J. Sci.
1958 Fd. Agr., 9 : 677-692.

368. RAMCHANDRA RAO, Y. and DUTT, A. The pests of the date palm in Iraq.
1921 Dept. Agr. Mesopotamia (Basrah), Mem., 6 : 1-21.

369. RANNOU, A. Financement des travaux nécessaires pour arriver à
1935 l'irrigation rationnelle des étendues cultivées de la région de Touggourt et des possibilités d'extension. Compte rendu général des Journées du Dattier, Biskra-Touggourt, 13-17. XI. 33, pp. 67-81. OFALAC, Alger.

370. RAO, Y.R. and DUTT, A. The pests of the date palm in Iraq. Dept. Agr.
1922 Mesopotamia. Memoir no. 6. Times Press, Bombay.

371. RAPPOPORT, Z. Notes on date palm growing in Israel. 1st FAO Int. Tech.
1959 Mtg. on Date Prod. and Proc. Tripoli, Libya. Paper no. 6A.

372. RAYNER, R.W. "Inflorescence rot" of date palms. Commonwealth Phyto-
1962 pathological News. Kew, London.

373. RAZ, D. The use of polyethylene wraps in the rooting of high date
1959 offshoots. Ann. Rep. Date Growers' Inst., 36 : 19.

374. REAL, P. Les myelois parasites des dattes (Lepid-Phycitinae). Rev.
1948 Path. Vég. Entomol. Agr. France, 27 : 59-64.

375. REAM, C.L. Date palm breeding - a progress report. Ann. Rep. Date
1975 Growers' Inst., 52 : 8-9.

376. REAM, C.L. and CARPENTER, J.B. Date palm breeding in California. 3rd FAO
1975 Tech. Conf. on Imp. Date Prod., Proc. and Mktg. Paper no. A.2.

377. REED, R. Growing and marketing dates with low overhead costs. Ann.
1924 Rep. Date Growers' Inst., 1 : 9-10.

378. REME, M. Méthode d'exploitation des eaux souterrains. Compte rendu
1935 général des Journées du Dattier, Biskra-Touggourt, 13-17. XI. 33, pp. 47-66. OFALAC, Alger.

379. REUTHER, W. and CRAWFORD, C.L. The effect of temperature and bagging on
1946 fruit set of dates. Ann. Rep. Date Growers' Inst., 23 : 3-7.

380. REUVENI, O. Observations on and experiments in the date palm groves in
1964 the Arava - II. (in Hebrew, summary in English). Nat. and Univ. Inst. of Agric., Rehovot.

381. REUVENI, O. Trickle irrigation of date palms. Ann. Rep. Date Growers'
1971 Inst., 48 : 16-18.

382. REUVENI, O. Drip versus sprinkler irrigation of date palms. Ann. Rep.
1974 Date Growers' Inst., 51 : 3-5.

383. REUVENI, O., ADATO, Y. and LILIEN-KIPNIS, H. A study of new and rapid
1972 methods for the vegetative propagation of date palms. Ann. Rep. Date Growers' Inst., 49 : 17-24.

384. RICHARDS, L.A. *et al.* Diagnosis and improvement of saline and alkali soils
1954 Agriculture Handbook no. 60, USDA, Washington, D.C.

385. RICHARDSON, A.M. The Bayford Park date grove. Queensland Ag. J., 75 (5) :
1952 253-262 (and Ag. Leaflet no. 259), Brisbane.

386. RICHARDSON, H.B. Summary of the date growers' report. Ann. Rep. Date
1941 Growers' Inst., 18 : 29-30.

387. RICHARDSON, H.B. The date palm in Cyprus. Ann. Rep. Date Growers' Inst.,
1964 41 : 2-3.

388. RIEUF, P. Contribution à l'étude du charbon de la datte *Aspergillus*
1963 *phoenicis* (Cda.) Thom. Al Awamia, 6 : 1-16.

389. RIVIERE, C. and LECQ, H. Traité pratique d'agriculture pour le nord de
1914 l'Afrique. Challamel, Paris.

390. ROBERTSON, PROWSCHOWSKY, A. Les palmiers sur la Côte d'Azur: leur resistance
1907 au froid. Bull. de la Soc. Nat. d'Acclimatation de France
(Revue des Sciences Naturelles Appliquées), 40 pages. Paris.

391. ROLLAND, G. Hydrologie du Sahara algérien (Chemin de fer transaharien),
1894 Impr. Nat., Paris.

392. ROSANIA, A. and MOSELE, L. La palma da dattero a Elche (Spagna). Riv. Agric.
1967 Subtrop., 61 : 15-20.

393. ROSE, Com.t. La culture du dattier dans le Sud Constantinois. Fontana,
1898 Alger.

394. RUNGS L'entomologiste, II : 8.
1946

395. SAAIDI, M., DUVAUCHELLE, S. and TOUTAIN, G. Etude de quelques facteurs
1975 conditionnant la réprise végétative des rejets de palmier-
dattier. 3rd FAO Tech. Conf. on Imp. Date Prod., Proc. and
Mktg. Paper no. B.12.

396. SAAIDI, M. and RODET, J. Lutte contre le bayoud. Efficacité de deux
1975 fongicides sur *Fusarium oxysporum* F. sp. *albedinis* "*in vitro*"
agent du bayoud. 3rd FAO Tech. Conf. on Imp. Date Prod.,
Proc. and Mktg. Paper no. B.11.

397. SACHS, G. Sur la présence d'*Omphalia* sp. Bliss dans une palmeraie
1967 mauritanienne. Fruits d'Outre Mer, 22 : 497-501.

398. SCHEIL, V. De l'exploitation des dattiers dans l'ancienne Babylonie.
1913 Revue d'Assyriologie, 10 : 1-9.

399. SCHOONOVER, W.R. Some suggestions on soil management in date gardens. Ann.
1935 Rep. Date Growers' Inst., 12 : 4-6.

400. SCHROEDER, C.A. Tissue culture of date shoots and seedlings. Ann. Rep.
1970 Date Growers' Inst., 47 : 25-27.

401. SELTZER, P. 1946 — Le climat d'Algérie. Inst. de Mét. Univ. d'Alger. Alger.

402. SHAFIK, M. and HILMY, A. 1939 — A mud brick oven for drying dates and controlling Ephestia and Myelois on dates. Bull. Soc. Fouad 1er Ent., XXIII : 233.

403. SHAMBLIN, A.J. 1924 — Eradication and control of date scale. Ann. Rep. Date Growers' Inst., 1 : 13, 14.

404. SHAMEL, A.D. 1930 — A bud variation in the Deglet Noor date palm. J. Heredity, 21 (4) : 164-166.

405. SHOWERS, B.J. 1927 — Soil management in light of the Rothampsted experiments. Ann. Rep. Date Growers' Inst., 4 : 2-5.

406. SINCLAIR, W.B., BARTHOLOMEW, E.T. and BLISS, D.E. 1941 — Composition of dates as affected by soil fertilizer treatments. Ann. Rep. Date Growers' Inst., 18 : 11-16.

407. SINGH, K. 1931 — Contribution towards our knowledge of the Aleurodidae of India. Mem. Dept. Agric. India, Ent. Ser., XII : 93. Calcutta.

408. SIRET, L. 1909 — Les Cassitérides et l'empire colonial des Phéniciens. L'Anthropologie, XX : 292. Paris.

409. SMEAD, P.F. 1965a — Mechanization of the date industry. 2nd FAO Tech. Conf. on Imp. Date Prod. and Proc., Paper no. 5.

410. SMEAD, P.F. 1965b — Storage and fumigation. 2nd FAO Tech. Conf. on Imp. Date Prod. and Proc., Paper no. 6.

411. SMIRNOFF, W.A. 1957 — La cochenille du palmier-dattier (Partlatoria blanchardi, Targ.) en Afrique du Nord. Comportement, importance économique, prédateurs et lutte biologique. Entomophaga, 2 : 1-99.

412. SMITH, H. 1928 — Fertilizing experiments. Ann. Rep. Date Growers' Inst., 5 : 2-4.

413. STICKNEY, F.A., BARNES, D.A. and SIMMONS, P. 1950 — Date palm insects in the United States. 846. USDA, Washington, D.C.

414. STOUT, A.B. 1924 — The viability of date pollen. J.N.Y. Bot. Gard., 25 : 101-106. Illus.

415. STREETS, R.B. 1933 — Heart rot of the date palm. Tech. Bull. Ariz. Agric. Exp. Sta., 48 : 443-469.

416. SURCOUF, J.M.R. 1922 — Recherches sur la biologie du Phoenix dactylifera. Bull. Soc. Hist. Nat. Af. Nord, 13 : 262-273, 293-312.

417. SURCOUF, J.M.R. 1923 Deuxième note sur les conditions biologiques du Stomoxys calcitrans, L. Bull. Mus. Hist. Nat., Paris, 2 : 168.

418. SWANN, G. 1925 Result of cover cropping. Ann. Rep. Date Growers' Inst., 2 : 6.

419. SWINGLE, W.T. 1904 The date palm and its utilization in the southwestern States. Bur. of Plant Industry, Bull. no. 53, USDA, Washington, D.C.

420. SWINGLE, W.T. 1928 Metaxenia in the date palm. J. Heredity, 19 : 257-268.

421. TALHOUK, A.M.S. Report on the diseases and insect pests of crops found in the Eastern province of Saudi Arabia. Arabian American Oil Co., Dhahran. Middle East Export Press Inc., Beirut, H.D.

422. TAQIZAADEH, F. 1950 Hardak: Funambulus palmarum, L. Entomologie et phytopathologie appliquées, 10 : 31-33 (Persian), 21 (French summary). Dept. Agric. Res., Min. Ag., Teheran.

423. TASCHDJIAN, E. Note sur les dattiers de la côte française des somalis. Rev. Bot. Appl., 13 : 798-801.

424. TEISSEIRE, P. 1959 Aperçu sur les ennemis et maladies du palmier-dattier dans les départements français d'Algérie et du Sahara et moyens mis en oeuvre pour les combattre. 1st FAO Int. Tech. Mtg. on Date Prod. and Proc. Tripoli, Libya. Paper no. 16.

425. TOURNEUR, J.C. 1975 Lutte contre les parasites et maladies - la lutte biologique menée par l'Institut Français de Recherches Fruitières Outre Mer (IFAC) contre Parlatoria blanchardi, Targ., ravageur du palmier-dattier. 3rd FAO Tech. Conf. on Imp. Date Prod., Proc. and Mktg., Paper no. C.35.

426. TOURNER, J.C. 1975 Lutte contre les parasites et maladies - expérimentation et lutte contre l'acarien du dattier Oligonychus afrasiaticus en Mauritanie. 3rd FAO Tech. Conf. on Imp. Date Prod., Proc. and Mktg., Paper no. C.36.

427. TOURNEUR, J.C., 1965 SACHS and LAUDEHO. Etudes sur Parlatoria blanchardi et ses prédateurs dans les palmeraies en Adrar (République Islamique de Mauritanie). 2nd FAO Tech. Conf. on Imp. Date Prod. and Proc., Paper no. 16.

428. TOUTAIN, G. 1965a Note sur l'épidémiologie du bayoud en Afrique du Nord. 2nd FAO Tech. Conf. on Imp. Date Prod. and Proc., Paper no. 10.

429. TOUTAIN, G. 1965b Essais de comparaison de la résistance au bayoud des variétés marocaines de palmier-dattier. 2nd FAO Tech. Conf. on Imp. Date Prod. and Proc., Paper no. 13.

430. TOUTAIN, G. 1965c Note sur l'épidémiologie du bayoud en Afrique du Nord. Al Alwamia, no. 15, pp. 37-45.

431. TOUTAIN, G. 1968 Essais de comparaison de la résistance au bayoud des variétés de palmier-dattier. II. Notes sur l'expérimentation en cours concernant les variétés marocaines et tunisiennes. Al Alwamia, no. 27, pp. 75-78.

432. TOUTAIN, G. 1975 Lutte contre le bayoud III. Frein à la propagation du bayoud en palmeraie. 3rd FAO Tech. Conf. on Imp. Date Prod., Proc. and Mktg., Paper no. B.10.

433. TOUTAIN, G., BACHRA, A. and MEUNIER, J.M. 1975 Fertilisation du palmier-dattier. Essai de fumure indicatif sur palmiers dattiers. 3rd FAO Tech. Conf. on Imp. Date Prod., Proc. and Mktg., Paper no. B.14.

434. TRABUT, L. 1912 Sur une maladie du dattier, le khamedj ou pourriture du régime. C.R. Acad. Sci., Paris, 154 : 304-305.

435. TRAUB, H.P. and ROBINSON, T.R. 1937 Date. pp. 26-35 in Improvement of subtropical fruits other than citrus. USDA Yearbook, Separate 1589.

436. TUNIS 1952 Service météorologique de la Tunisie. Normales et statistiques diverses. Tunis.

437. TUNIS 1953 Service tunisien des statistiques. Annuaire statistique de la Tunisie, 1951-1952. Tunis.

438. TYLER, J. 1941 Plants reported resistant or tolerant to root knot nematode infestation. USDA Misc. Publ. no. 406. 91 pages.

439. UNITED STATES DEPARTMENT OF AGRICULTURE. 1962 Quarantine. Nursery stock, plants and seeds. Title 7, Agriculture, chapter III, Agricultural Research Service, Department of Agriculture, part 319. Foreign Quarantine Notices, Section 319.37, p. 19. Washington, D.C.

440. UVAROV, B.P. 1928 Locusts and grasshoppers. London, Imperial Bureau of Entomology. 352 pages.

441. UVAROV, E.B. and CHAPMAN, D.E. 1958 A dictionary of science. Penguin, London.

442. UYTTENBOOGAART, D.L. 1927 Some remarks regarding the discovery and biology of Dactylotrypes uyttenboogaarti, Eggers. Tijdachr. Ent., 70 : 40.

443. VEIHMEYER, F.J. and HENDRICKSON, A.H. 1960 Essentials of irrigation and cultivation of orchards. Extn. Serv. Circ. Calif. Agric. Exp. Sta. no. 486.

444. VIENNOT-BOURGIN, G. 1949 Les champignons parasites des plantes cultivées. Paris. 2 Vol., Masson et Cie., editeurs. 1851.

445. VINCENT, L.E. and LINDGREN, D.L. 1958 Control of the date mite Oligonychus pratchsis, Banks, in California. Ann. Rep. Date Growers' Inst., 35 : 15-17.

446. VINSON, A.E. 1941 The effect of climate conditions on the rate of growth of date palms. The Botanical Gazzette, 7 (4) : 324-327.

447. VIS, E.G., BROWN, G.K. and PERKINS, R.M. 1972 Date pollination made easier. Agricultural Research, USDA, 20 (7) : 15.

448. VIS, E.G., PERKINS, R.M. and BROWN, G.K. 1971 Mechanical pollination experiments with Deglet Noor date palms in 1970. Ann. Rep. Date Growers' Inst., 48 : 19-22. USDA, Riverside, California.

449. WAKSMAN, S.A. 1938 Humus. Williams and Wilkins, New York, 2nd ed.

450. WALKER, J. and MITCHELL, D.H. 1944 The fumigation of dates. Ann. Rep. Date Growers' Inst., 21 : 4-6.

451. WEISS, A. 1912 Sur un diptère du genre nuisible à l'état de larva à la culture du dattier dans l'Afrique du Nord. Bul. Soc. Hort. Nat. Afr. Nord, IV (3) : 68-69.

452. WERTHEIMER, M. 1956a Essais de limitation par ciselage de la production du palmier-dattier Deglet Noor dans le Ziban (région de Biskra). Fruits, 11 : 429-433. Paris.

453. WERTHEIMER, M. 1956b Recherches et observations sur la plantation des "rejets" des palmiers dattiers dans le Ziban (région de Biskra). Fruits, 11 : 481-487.

454. WERTHEIMER, M. 1957a La pollination du palmier-dattier (Phoenix dactylifera). Fruits, 12 : 305-313. Paris.

455. WERTHEIMER, M. 1957b Problèmes du drainage dans les palmeraies du sud constantinois. Fruits, 12 : 345-352. Paris.

456. WERTHEIMER, M. 1958 Un des principaux parasites du palmier-dattier algérien le Myelois decolor. Fruits, 13 : 309-323.

457. WHITAKER'S ALMANACK. 1965 J. Whitaker & Sons Ltd. London.

458. WHITTLESEY, H.R. 1933 Ripening dates earlier by using different pollen. Ann. Rep. Date Growers' Inst., 10 : 9.

459. WHYTE, R.O. and TRUMBLE, H.C. 1953 Legumes in agriculture. FAO Agricultural Study no. 21.

460. WILLAR 1927 Proc. Hawaii Ent. Soc., 6 : 372.

461. WILLCOCKS, F.C. 1922 A survey of the more important economic insects and mites of Egypt. Bull. Sult. Agric. Soc., Tech. Sec., no. 1. Cairo.

462. WILTSHIRE, E.P. 1957 The Lepidoptera of Iraq. 2nd ed. London, Nicholas Kaye. 162 pages.

463. WINDER, G. 1968 — Production and marketing of Medjool dates in Bard, California. Ann. Rep. Date Growers' Inst., 45 : 10-11. Coachella.

464. WINKLER, H. 1843 — Dattelpalme. (In SCHMIDT, G.A. and MARCUS, A. Handbuch der trop. und subtrop. Landwirtschaft, 2 : 118). Mittler Verl., Berlin.

465. WINSLOW, M.M. 1928 — Date fertilizer trials in the Coachella Valley. Ann. Rep. Date Growers' Inst., 5 : 5.

466. WRIGHT, J.F. 1975 — Mechanization of date cultural practices. Ann. Rep. Date Growers' Inst., 52 : 34.

467. WOLCOTT, G.N. 1936 — Insectae borinquenses. J. Agric. Univ. Puerto-Rico, XX : 1. Puerto-Rico.

468. YARMELL, S.H. 1942 — Influence of the environment on the expression of hereditary factors in relation to plant breeding. Science, 96 : 505-508.

469. ZAMBEU — (Note on boring beetles). Rev. Entom., XIX : 13.